Carbon Nanotubes Handbook

Carbon Nanotubes Handbook

Edited by **Lindy Bowman**

New York

Published by NY Research Press,
23 West, 55th Street, Suite 816,
New York, NY 10019, USA
www.nyresearchpress.com

Carbon Nanotubes Handbook
Edited by Lindy Bowman

International Standard Book Number: 978-1-63238-070-8 (Hardback)

Contents

Preface

This book has been a concerted effort by a group of academicians, researchers and scientists, who have contributed their research works for the realization of the book. This book has materialized in the wake of emerging advancements and innovations in this field. Therefore, the need of the hour was to compile all the required researches and disseminate the knowledge to a broad spectrum of people comprising of students, researchers and specialists of the field.

Carbon nanotubes are large molecules of pure carbon that are long and thin and are shaped like tubes of nanometer scale diameter with quasi-one-dimensional structure. In the past 20 years, carbon nanotubes have attracted a lot of attention from chemists, electronic device engineers, physicists, and material scientists, due to their fine optical, mechanical, structural, chemical and electronic properties. The demand for innovative industrial applications of carbon nanotubes is growing significantly. This book encompasses latest research topics about the synthesis technologies of carbon nanotubes and nanotube-based composites, along with their applications such as Electrical and Biomedical Applications and Carbon Nanotubes for Green Technologies. It will serve as a helpful source of information for engineers, researchers and students.

At the end of the preface, I would like to thank the authors for their brilliant chapters and the publisher for guiding us all-through the making of the book till its final stage. Also, I would like to thank my family for providing the support and encouragement throughout my academic career and research projects.

Editor

Electrical and Biomedical Applications of Carbon Nanotubes

Carbon Nanotube Transparent Electrode

Jing Sun and Ranran Wang

Additional information is available at the end of the chapter

1. Introduction

In the modern world, transparent conductive films (TCF) are extremely common and critically important in electrical devices. In our homes or offices, they are found in flat panel displays such as in TVs, laptops and in touch panels, of phones, tablet computers, E-readers and digital cameras [1]. Besides, they are also used as the electrodes for photovotaic devices such as solar cells [2] and organic light-emitting diodes (OLEDs) [3]. Liquid crystal display (LCD) is by far the largest user of transparent conductive films but many devices are showing rapid growth in popularity such as touch panels (362 million units in 2010 with annual growth of 20% through 2013), E-paper (30 fold growth expected from 2008 to 2014), and thin film solar cells (expected sales of over $13 billion by 2017) [4].

The dominant transparent conductive material used today is tin doped indium oxide (ITO) with a demand growing at 20% per annum [5]. ITO has been studied and refined for over 70 years, and as a result, the material offers many beneficial properties. However, ITO has certain drawbacks, mainly reflected on the depleted supply of raw material and their brittleness. The supply of indium is constrained by both mining and geo-political issues, which leads to dramatic price fluctuations over the last decades, from $ 100- $ 900, as shown in Figure 1. The high price of indium determined the high cost of ITO, since they compose nearly 75wt % of a typical ITO film [6]. In addition to the raw materials, the expense of setting up and maintaining a sputtering deposition line, as well as the low deposition yield (3-30%) [7] also increases the cost of ITO. Though current devices are typically based on rigid substrates, there is a continued trend toward flexible devices. As ITO tend to fracture at strains of 2%, they are completely unsuitable for using in flexible electronics. Therefore, new transparent electrode materials have rapidly emerged in recent years, including carbon nanotubes (CNTs), graphene and metal nanowires. The intrinsically high conductivity coupled with high aspect ratio yields films with high transmittance, adequately low sheet resistance, and superior mechanical flexibility. These material properties, combined with inexpensive material and deposition costs make these emerging nanomaterials very attractive for as transparent electrodes. Of the three dominant nanoscale materials, CNTs are perhaps the most promising and mature intensively investigated.

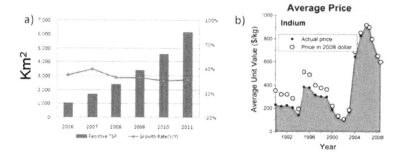

Figure 1. (a) Global demand for resistive style touch panels by area; (b) Average price of Indium over the last several decades; Reprinted with permission from reference [4] copyright 2011 Wiley.

This review will focus on transparent electrode made of CNTs, and six main parts will be covered.

1. At first, some basic theories and parameters for characterizing transparent conductive materials will be presented so that the following parts of the review can be profoundly understood.

2. CNTs prepared from different methods or modified under various conditions have diverse physical and chemical properties, which will yield films with distinct performance. Therefore, in the second part, CNTs of different types will be investigated, and the performance of the as prepared thin films will be compared.

3. One of the major advantages in using CNTs is their ability to be applied to substrates from solution, which opens up many alternative deposition techniques. Therefore, one of the primary research areas for making transparent conductive films is to process the CNT material into printable inks.The third part will outline major approaches to disperse CNTs and focus on the most important details with regards to making transparent conductive films.

4. In the fourth part, a variety of techniques for making transparent conductive CNT films will be presented and evaluated.

5. During the solubilization step, non-conducting dispersants are induced, which sacrifice the conductance of the films a lot. Therefore, post-treatment needs to be done to remove them for enhancing the performance of the films. In the fifth part, various methods used to improve the performance of the transparent conductive films after their preparation will be discussed.

Finally, the latest progress on CNT transparent conductive films and their applications on electrical devices will be summarized.

2. Optoelectronic properties

The two most important features for a transparent conductingmaterial are its sheet resistance (Rs) and optical transparency. The sheet resistance is defined as Rs = R(W/L), where R is DC resistance, W and L are width and length of the film. Grüner et al. [8] developed a suitable merit, the DC conductivity/optical conductivity (σ_{dc}/σ_{op}), to compare the performance of various transparent conductors based on the standard percolation theory, in which each bundle of nanotubes was counted as one conducting stick. They assumed the conductivity ratio σ_{dc}/σ_{op} remains constant for nanotube networks with different densities in the measured optical frequency range. By plotting RsvsT and fitting the data to equation 1, one can estimate the value ofσ_{dc}/σ_{op}. This value is often used as a Figure of Merit for transparent conductors since high values of σ_{dc}/σ_{op} leads to films with high T and low Rs.

$$T = \frac{1}{1 + \frac{2\pi\sigma_{op}}{cR_s\sigma_{dc}}} \tag{1}$$

Geng et al. [9] found that this equation can be fitted well to the curve of single-walled carbon nanotube TCFs, nevertheless can not be fitted well with carbon nanotubes of other types. They modified the equation as follows:

$$T = t \bullet \left(1 + \frac{188\ (\Omega)}{R_s} \frac{\sigma_{op}}{\sigma_{dc}}\right)^{-2} \tag{2}$$

The parameter t may represent the optical property of CNT films. A high t value gives a high transmittance for the CNT films. The t value of SWCNT films is 0.999, while that of MWCNT is much lower, around 0.884.

Recently, Coleman et al. [5] modified this model to evaluate thinner (more transparent) films. They found that the data tend to deviate severely from the fits for thinner films, as seen from Figure 2. This deviation has been observed before [10-12] and tends to occur for films with T between 50% and 92%.Thus,σ_{dc}/σ_{ac} fails to describe the relationship between T and Rsin the relevant regime. The deviation from bulk-like behavior as described in Equation1, can be explained by percolation effects [13]. Such effects become important for very sparse networks of nano conductors. When the number of nanoconductors per unit isvery low, a continuous conducting path from one side of the sample to the other will generally not exist.

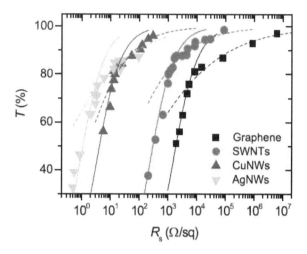

Figure 2. Typical graph of transmittance (generally measured at 550 nm) plotted versus sheet resistance for thin films of nanostructured materials. Reprinted with permission from reference [5] copyright MRS.

As more nanoconductors are added, at some point (the percolation threshold) the first conducting path will be formed. As more material is added, more conductive paths are formed, and the conductivity of the network increases rapidly. Eventually it reached a "bulk-like" value above which it remains constant. Percolation theory describes how the dc conductivity of sparse networks depends on network thickness and predicts a non-linear, power law dependence:

$$\sigma_{dc} \propto (t - t_c)^n \tag{3}$$

where tis the estimated thickness of the network, t_c is the thickness associated with the percolation thres hold, and nis the percolation exponent. This leads to a new relationship between T and Rs, which applies to thin, transparent networks:

$$T = \left[1 + \frac{1}{\Pi}\left(\frac{Z_0}{R_s}\right)^{1/(n+1)}\right]^{-2} \tag{4}$$

where Π is the percolative FoM:

$$\Pi = 2\left[\frac{\sigma_{dc}/\sigma_{op}}{(Z_0 t_{min}\sigma_{op})^n}\right]^{1/(n+1)} \tag{5}$$

Here, t_{min} is the thickness below which the dc conductivity becomes thickness dependent. It scales closely with the nanostructures' smallest demision, tmin ≈ 2.33 D. The high Tportion of data in Figure 1 was fitted using Equation 4, and good fits allow the calculation of both n and Π. Analysis of these equations shows that large values of Π but low values of n are desirable to achieve low Rs coupled with high T, which are used to evaluate the performance of CNT films with high performance.

In addition to their sheet resistance and optical transparency, the stability and mechanical durability are also critical criteria to evaluate the performance of transparent conductors. Undoped CNT films exhibit excellent stability upon exposure to atmospheric conditions, as seen in Figure 3 [14]. Doping with nitric acid or $SOCl_2$ could decrease the sheet resistance significantly, however at the expense of sacrificing their stability [15-17]. The sheet resistance of undoped SWCNT films decreases slightly with increasing temperature, which is consistent with the electrical behavior of semiconductors. Thermal stability of doped CNTs is dependent on dopants since elevated temperatures may increase chemical reactions or enhance the desorption of dopants out of the films. CNT-PET thin films are significantly more flexible than commercial ITO/PET films. They can be bent all the way to 180° without a significant change in resistance, [18] and the conductivity of the films can be retained after 500 bending cycles [19].

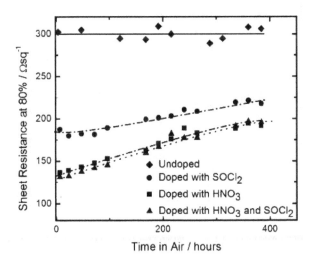

Figure 3. Absolute sheet resistance versus time in air of four SWNT films. Reprinted with permission from reference [14] copyright Wiley.

3. The choice of Carbon Nanotubes

Carbon nanotubes synthesized from different methods or processes have diverse material qualities, such as the degree of purity, the defects, their length and diameters, and the chiral-

ities, which are presumably important factors in determining the film performance.There-
fore, the choice of CNTs as well as their further treatment is markedly important. Young
Hee Lee group [9,20] did systematical analysis to investigate the CNT quality dependence.
In their work, single-walled carbon nanotubes (SWCNT), double-walled carbon nanotubes
(DWCNT), thin multiwalled carbon nanotubes (t-MWCNT) and multiwalled carbon nanotubes (MWCNT) powders were separately dispersed in deionized water with sodium dodecyl sulfate (SDS) and dichloroethane (DCE) by sonication and sprayed onto poly (ethylene
terephthalate) (PET) substrates to fabricate thin films. The sheet resistance and transmittance
of each film was measured and compared. As seen in Figure 4, the film's performance
changes dramatically for different types of CNTs dispersed in deionized water with SDS, as
well as in DCE. The TCFs fabricated with SWCNTs show the best film performance among
all the selected CNTs. The trends of film performances are similar for the TCFs fabricated by
using the CNT solution dispersed in deionized water and in DCE, which is
SWCNTTCF>DWCNTTCF> t-MWCNTTCF>MWCNTTCF. Furthermore, they analyzed the
defects and metallicity by Raman spectra, and found that CNTs with fewer defects and high
content of metallic tubes leads to TCFs with higher conductivity. Nevertheless, in Li's report, [21]. MWCNTTCFs exhibit better performance than SWCNTTCFs. They indicated that
MWCNT have more conductive π channels than SWCNTs does, therefore MWCNTs have
better electronic transportability. In the case of a MWCNT where conduction occurs through
the outer most shell, the large diameter of the outernanotube causes the gap to approach 0
eV and the nanotubeto become basically metallic. On the contrary, 2/3 of SWCNTs are semiconducting. The other reason they mentioned is that the MWCNTs they used are longer
than SWCNTs, which could decrease the contacts numbers. Another point needs to be addressed is that dimethylformamide (DMF) which was chosen as the solvent in their work is
actually not efficient to exfoliate SWCNTs. Therefore, SWCNTs bundled together which
would open up an energy gap or pseudo gap owing to intertube interactions. We believe
this is a critical reason for the worse performance of SWCNTTCFs in their work.

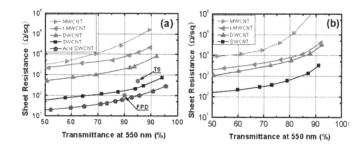

Figure 4. Characteristic sheet resistance-transmittance curves for various CNT-films. Each curve contains several data
points from films with different numbers of sprays by a CNT solution dispersed in (a) deionized water with SDS and (b)
DCE without dispersant. Reprinted with permission from reference [9].

SWCNTs synthesized by different methods such as arc discharge (Arc), catalytic chemical vapor deposition (CVD), high pressure carbon monoxide (Hipco), and laser ablation (Laser) were also analyzed systematically [20]. After the SWCNT powder was characterized, each of them was dispersed in deionized water with sodiumdodecyl sulfate (SDS) by sonication followed by aspray process to fabricate the SWCNT film onto PET substrates.By analyzing the SWCNT film performance varying with the SWCNT parameters, they found that the metallicity of the SWCNTsextracted from G'-band intensity of Raman spectros copy and the degree of dispersion in the solutionare the most decisive factors in determining the film performance. Figure 5 shows that the film performance changes dramatically with different types of SWCNTs. The TCFs fabricated with Arc SWCNTs result in the best film performance, consistent with previous report [22]. The sheet resistance of the Arc TCF is ~160Ω/sq at a transmittance of 80%, which can be used in a wide range of applications from touch panels to electrodes for future flexible displays.

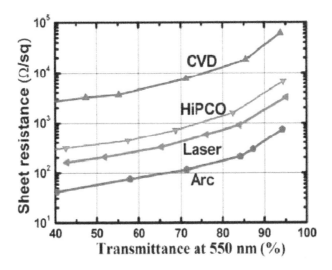

Figure 5. Characteristic curves of sheet resistance-transmittance of TCFs fabricated by various SWCNTs. Reprinted with permission from Ref. [20].

In order to investigate the underlying reason, CNTs were characterized with SEM, TEM, TGA and Rama spectra. TEM analysis showed that the diameter of individual nanotube synthesized with CVD and Hipco process were about 1nm, smaller than those (~1.4 nm) of Laser and Arc SWCNTs. The CVDSWCNTs had the smallest average bundle size, as estimated from the SEM images, where as the Laser sample exhibited the largest average bundle size among samples. Carbonaceous particleson the SWCNT bundles are present in the CVDSWCNTs. The Arc SWCNTs have relatively well-defined crystallinity without amorphous carbonson the tube walls, although the bundle size of the Arc sample is smaller than that of the Laser sample. Figure 6 disclosed that the influence of the purity of the SWCNT is less deterministic, particularlyin CVD and HiPCOSWCNTs, where as the diameter has a strong correlation to the sheet conductance of SWCNT film. The sheet conductance of the film increases consistently with increasing diameters of nanotubes, as shown in Figure 6. This can be attributed to the decreasing band gap with increasing diameters of semi-conducting SWCNTs. Although individual metallic tubes are independent of the diameters, there are usually a pseudogap induced by tube-tube interactions, which is also inversely proportional to thetube diameter. Thus, the conductivityof the metallic nanotubes reveals the similar diameter dependence to semiconducting ones.

The radical breathing modes (RBM) of Raman spectra were used to characterize the metallicity of SWCNTs [20]. At 514 nm, the Laserand Arc SWCNTs reveal the semiconducting behavior exclusively, on the other hand, CVD and HiPCOSWCNTs containboth metallic and semiconducting nanotubes. At 633 nm, the Laser and Arc SWCNTs pick up mostly metallic SWCNTs, where as the CVDSWCNTs retain mostly semiconducting properties (less prominent Fano line) and the HiPCOSWCNTscontain both the metallic and the semiconducting behaviors. Other than RBM mode, the G'-band intensity is strongly correlated with the metallicity of SWCNTs. Despite the abundance of metallicity, the presence of defects on then anotube walls that may act as scattering centersdegrades the conductivity of the SWCNT network [23]. The intensity of the D-band indicates the amount of defects on the nanotube walls. Therefore, anappropriate parameter to express conductivity of nanotubes for SWCNTs is the intensity ratio, G'-band/D-band. High content of metallicity and few defects on the nanotube walls will be desired for high conductivity of the SWCNT films.

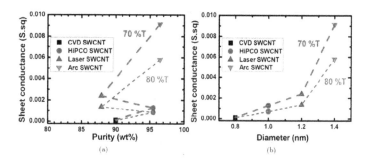

Figure 6. The sheet conductance of TCFs at transmittance of 70% and 80% versus (a) purity and (b) diameter of SWCNT powders. Reprinted with permission from Ref. [20].

The purity affects the conductivity.The diameter contributes to the conductivity via bandgap described in the previous paragraph. More defects reduce the *mean free path* of carriers and decrease the mobility of carriers in nanotubes.The conductivity is proportional to the metallicity of nanotubes and inversely proportional to the number of scattering centers or defects [24-26]. Considering all these factors, a material quality factor Q_m was defined to govern the conductivity of SWCNTs:

$$Q_m = P \times \left(e^{E_{pq}/2k_BT} \times \overline{\sum I_M} \right) + e^{-E_q/2k_BT} \times e^{(E_i - E_f)/k_BT} \times \overline{\sum I_S} \qquad (6)$$

where Eg = 0.82/D (eV), Epg = 0.105/D (eV), D is the average diameter of individual SWCNTs, P is the purity of the sample, E_i is the intrinsic Fermi Level, E_f is the Fermi Level for the extrinsic semiconductors, k_B is the Boltzmann constant and T is the temperature of the system. Here I_S (I_M) is defined as

$$I_S(I_M) = I_{G'/D} \times \frac{A_S(A_M)}{A_M + A_S} \qquad (7)$$

where $A_S(A_M)$ is the areal intensity of semiconducting (metallic) peaks of RBMs from Raman shift. After calculation, it was observed that the sheet conductance reveals a linear relationship with the material quality factor. Although this empirical formula is not rigorous, it can provide atleast a means to estimate material quality that governs the conductivity of the SWCNTTCFs. Forinstance, large diameter, higher purity, less defects (lower intensity of D-band), and more metallic nanotubes (higher intensity of G'-band) will give better conductivity of the SWCNTTC. From this point of view, the Arc TCF is the best sample providing the highest conductivity in comparison toTCFs made by other types of SWCNTs.

In addition to the material parameters discussed above, the length of SWCNTs is also crucial to the TCF performance. According to the percolation theory, a conducting path could be formed at a lower density for longer nanotubes, which means at the same sheet resistance, TCFs prepared with longer nanotubes should exhibit higher optical transparency. This conjecture has been confirmed by experiments [27,28]. In order to optimize the CNTs quality, such as their purity, their dispersibility and the content of metallic tubes, some pretreatments need to be done. Several attempts have been tried to purify the CNT powders.Generally, Gas phase reaction or thermal annealing in air or oxygen atmosphere is used to remove amorphous carbon [29,30]. The key idea with these approaches is a selective oxidative etching processes, based on the fact that the etching rate of amorphous carbons is faster than that of CNTs. Since the edge of the CNTs can be etched away as well as carbonaceous particles during the annealing, itis crucial to have a keen control of annealing temperatures and annealing times to obtain high yield. Liquid-phasereactions in various acids are always conducted to remove the transitionmetal catalysts [31-33]. Hydrochloric acid, nitric acid and sulfuric acid are the most commonly used acid, and the purification effect is dependent on the concentration, the reaction temperature and the reaction time. In addition to their reaction with metal catalysts, nitric acid and sulfuric acid could induce some carboxyl or hydroxyl groups onto the walls of nanotubes, which will improve their dispersibility in water [34,35]. However, some damages were introduced during this process. Therefore, subsequent annealing or ammonium treatment was sometimes carried out to repair the wall structures of the nanotubes to fulfill some special requests.[36]. In order to enhance the content of metallic tubes, discriminated adsorption and separation or ion change chromatography was generally used.

4. CNT Ink Preparation

One of the major advantages in using CNTs overmore conventional metal oxides is their ability to be applied to substrates from solution, which opens up many alternative deposition techniques. Therefore, one of the primary areas of research for making transparent conductive films is finding ways to process the CNT materials into printableinks.The first part of the ink making process is in finding suitable ways to disperse the CNT materialinto solution. Commercial SWCNTs always aggregated into thick bundles due to their high surface energy and strong van der Waals force between tubes. However, the conductivity of the SWCNTTCFs is inversely proportional to the bundle size considering tube-tube junction resistance [37]. Therefore, it is crucial to exfoliate SWCNT thick bundles into thinner or even individual ones.

There are three major approaches to dispersing CNTs:

a.   dispersing CNTs in neat organic solvents [38,39];

b.   dispersing CNTs in aqueous media with the assistant of dispersing agents such as surfactants and biomolecules [40];

c.   introducing functional groups which will help draw the CNTs into solution [41].

Each of these methods have advantages and disadvantages in terms of making processable CNT based inks.

Direct solubilization of CNTs in a suitable solvent is perhaps the simplest and the most favorable method from a manufacturing point of view, since there are no solubilization agents involved which could create processing issues during manufacturing,and also lead to decreased conductivity in the as deposited film. A range of solvents have been tried to exfoliate SWCNTs, and exhibit tremendous differences on the efficiency. The major issue with using these organic solvents has beenthe inability to disperse CNTs at a concentration high enough to be useful for industrial applications (>0.1 g/L). Recently, workby Prof. Coleman's group [42] has shown that the solvent cyclohexylpyrrolidone (CHP) can disperse CNTs up to 3.5 g/L with high levels of individual tubes or small bundles and can keep stable for at least one month. However, the high boiling point of this solvent may be an issuein high speed roll-to-roll manufacturing on plastic. Continuing to search for optimal solvents which can disperse CNTs at high concentrations and have a reasonably low boiling point (150 °Cor below) could lead to a facile manufacturing process for high performance transparent conductive films.

Over the years, significant efforts have been devoted to finding a suitable parameter to guide the selection of good solvents. Three major theories have been proposed, which are non-hydrogen Lewis base theory, [43] polar π system and optimal geometry theory [44] and Hansen parameter [42]. According to non-hydrogen Lewis base theory, all of the solvents can be divided into three groups on the basis of their properties. Class 1consists of the best solvents, N-methylpyrrolidone (NMP),N,N-dimethylformamide (DMF), hexamethylphosphoramide(HMPA), cyclopentanone, tetramethylenesulfoxide andε-caprolactone (listed in decreasing order of optical densityof the dispersions), which readily disperse SWNTs, forminglight-grey, slightly scattering liquid phases. All ofthese solvents are characterized by high values for electron-pair donicityβ[45], negligible values for H-bond donation parameter α,[46] and high values for solvochromic parameter$\pi*$. Thus, *Lewis basicity* (availability of a free electronpair) without H-donors is key to good solvation of SWNTs.Class 2 contains the good solvents, toluene, 1,2-dimethylbenzene (DMB),CS_2, 1-methylnaphthalene, iodobenzene,CHCl3, bromobenzene and 1,2-DCB. They show $\alpha \approx \beta \approx 0$ and high valueof $\pi*$. Class 3 entails the badsolvents, n-hexane, ethylisothiocyanate, acrylonitrile, dimethyl sulfoxide (DMSO),water and 4-chloroanisole. Badsolvents would have $\alpha = \beta = \pi* \approx 0$. However, the high electron-pair donicity alone has proven tobe insufficient, as dimethyl sulfoxide (DMSO) is not an effectivesolvent for SWNTs even though it contains three lone pairs [47]. A systematic study of the efficiency of a series of amide solvents to disperse as-produced and purified laser-generated SWNTssuggested that the favorable interaction between SWNTs andalkyl amide solvents is attributable to the highly polar π systemand optimal geometries (appropriate bond lengths and bondangles) of the solvent structures [48]. However, this conclusion is some what undermined by the poor solubility of SWNTs intoluene [47]. Recently, Coleman et al found that the dispersibility of SWCNTs was intimately related with the Hansen parameters of the solvents and it is more sensitive to the dispersive Hansen parameter than thepolar or H-bonding Hansen parameter. The dispersion, polar, and hydrogen bonding Hansenparameter for the nanotubes is estimated to be$<\delta_D> = 17.8$ MPa$^{1/2}$,$<\delta_P> = 7.5$ MPa$^{1/2}$, and$<\delta_H> = 7.6$ MPa$^{1/2}$. Success ful solvents exist in only a small volume of Hansen space, which is $17 <\delta_D< 19$ MPa$^{1/2}$, $5 <\delta_P< 14$ MPa$^{1/2}$, $3 <\delta_H< 11$ MPa$^{1/2}$. Hansen parameters have been used successfully to aid solvent discovery. Unfortunately they are not perfect. A number of non-solvents exist in the region of Hansen parameter space close to the solubility parameters of nanotubes.

Compared with organic solvent, it is more efficient to exfoliate SWCNTs into thin bundles or even individual tubes with the assistant of dispersants. The most common dispersants used in TCFs are anionic surfactants including sodium dodecyl sulphate (SDS) and sodium dodecylbenzenesulphonate (SDBS). They are preferable dispersants because nanotubes can be highly exfoliated by them at rather high concentrations [49]. Besides, they nearly have no absorption over the visible spectrum region. However, they are not without disadvantage. Large amount of them is needed to exfoliate nanotubes into thin bundles; usually the CMC (critical micelle concentration) value should be reached [50]. Their residue will increase the sheet resistance of nanotube films significantly since they are nonconductive. In recent years, a lot of research has been done on the dispersion of CNTs with biomolecules such as DNA and RNA [51-54]. There are a number of advantages using them as dispersants.First, they can coat, separate, and solubilize CNTs more effectively with their phosphate backbones interacting with water and many bases binding to CNTs [55]. DNA wrapped around

CNTs helically and there were strong π-π interactions between them [56]. Charges were transferred from the bases of DNA to CNTs leading to the change of their electron structures and electrical property [57]. 1 mgDNA could disperse an equal amount of as-producedHiP-COCNT in 1 ml water, yielding 0.2 to 0.4 mg/ml CNT solution after removal of non-soluble material by centrifugation. Such a CNTsolution could be further concentrated by ten-fold to give a concentration as high as 4 mg/ml [52]. Jeynes's research disclosed that total cellular RNA showed better dispersion ability than dT(30) which was the most effective oligonucleotide dispersants in previous reports [54]. Second, the amount of DNA needed to exfoliate CNTs into thin bundles was much less than common surfactants such as SDS. In Zheng's work, the weight ratio between SWCNTs and DNA was 1:1 [52] while the dosage of RNA in Jeynes's work was lower, only half amount of the nanotubes [54]. By contrast, ten fold of SDS was needed to exfoliate SWCNTs efficiently [11,58]. High dosage of dispersant is not preferred since they are nonconductive and their residue will decrease the conductivity of the films significantly. Third, they have little absorption over the visible range and will not decrease the transmittance of CNT films. Last but not least, as biomolecules, they are easily degraded and removed by acid, base or appropriate enzyme. Jeynes et al [54] have used RNA to disperse CNTs and digested them by RNase effectively.

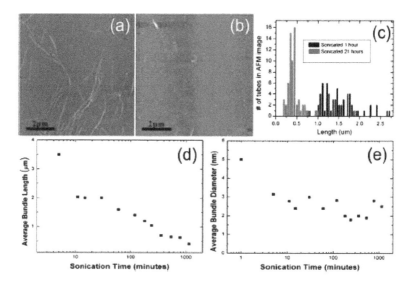

Figure 7. Effects of sonication on SWNT bundle length anddiameter. (a) and (b) AFMimage of SWNTs absorbed on a silicon waferafter (a) 1 h and (b) 21 h of sonication time. (c) Histogram of bundle length distribution taken from several AFM images for 1 h (black) and 21 h (red) of sonication. Plot of the (d) average bundle diameter and (e) average bundlelength for various sonication times measured from AFM images. Reprinted from Ref. [37] copyright AIP.

The final solubilization approach involves functionalizing CNT walls with covalently bonded molecules. The most commonly used process is introducing carboxyl groups by reacting with concentrated acid, such as nitric acid and sulfuric acid [59]. Although thismethod has been proven to lead to CNT solutions with high concentrations of thin bundles, the films made from these tubes tend to have extremely low conductivity values, as the functionalization procedure inducesdefects into the pristine CNTsp^2 bond structure.

For all solubilization approaches, energy must be imparted to the system to break the strong van der Wall force between tubes. This is commonly done by mixing techniques such as high-shear mixing, rotor-stator, three-roll milling, ball milling, homogenizers, and ultrasonication. Among these, ultrasonication is the most commonly used and the most efficient technique to prepare SWCNT water solution. The vibration of the sonicationtip in the solution causes pressure waves which expand and collapse dissolved gas in the liquid; the collapse of these bubbles causes temperature of local zones exceeding 10 000 °C, [60] which can impart enough energy to separate CNTs from each other, long enough for surfactants to surround the tubes and prevent them from aggregating. However, such high energy of sonication would introduce defects onto the walls of CNTs or even shorten them [37]. As seen from Figure 7, the diameter of the bundles decreases sharply from 5 to 3 nm in the first 5 min of sonication, and then remains 2-3 nm after that. However, the length of the tubes decreases exponentially with sonication time from 4μm initially, to 0.4μm after about 21 h of sonication. Therefore, suitable sonication powder and time needs to be chosen to make SWCNT inks with thin bundles and long length.

Figure 8. Freestanding SACNT film drawn out from a230-mm-high SACNT array on an 8-inch silicon wafer. The film in the visualfield is about 18cm wide and 30cm long. b) SEM image of the SACNT array on the silicon wafer in side view. c) SEM image of an SACNT film intop view. Reprinted with permission from Ref. [63] copyright Wiley

5. Film Fabrication

Many techniques have been developed to prepare CNT thin films, including both dry and solution-based methods. Although solution-based techniques are the mostly commonly used and industry preferred, dry method is negligible for preparing high performance TCFs. Direct growth of CNT films is one of the typical dry method. CVD can grow CNT films either randomly distributed or aligned by controlling the gas flow, catalyst patterns, or by using a substrate with a defined lattice structure [61]. Compared with a solution-based process, the direct growth method leads to films with individually separated tubes with fewer defects and better CNT-CNT contact, which leads to highly conductive films [62]. However, films directly grown on a substrate may have significant amounts of residual catalyst, imprecise density control, and substrate incompatibility for device integration. Furthermore,CVD is a high vacuum, high temperature process and is not compatible with substrates used in the emerging plastic electronics field.

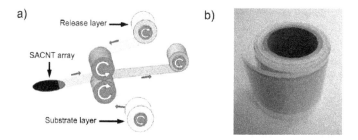

Figure 9. Production and performance of SACNTTCFs. a) Illustration of the roll-to-roll setup for producing composite TCFs. b) A reel of SACNT/PE composite TCF produced by the roll-to-rollsetup. The grey central region of the reel is the SACNT/PE composite TCF. Reprinted with permission from Ref. [63] copyright Wiley.

In 2002, a method was pioneered by Dr Fan's group [63] and involves drawing out MWCNT films directly fromas-grown super aligned CNT (SACNT) arrays. An example of such process and films are shown in Figure 8. An SACNT array is a special kind of vertically aligned MWCNT array having a higher surface density and better alignment of MWCNTs than an ordinary one.Typically, an SACNT array with an area of 0.01 m^2 can be totally converted to a SACNT film of ~6–10 m^2, depending on the height of the SACNT array. Unlike the solution-based process, an entire SACNT array can be converted to films without any significant loss by the drawing process, which will lower the cost. Another crucial advantage of this solution-free process is that it can be straight forwardly incorporated into a roll-to-roll process to make SACNT/polymer-sheet composite films. In a roll-to-roll process as shown in Figure 9a, aSACNT film is drawn out, then sandwiched by a release layer and a substrate layer, and pressed by two close rollers tightly, forming an SACNT/substrate composite film. The release layer, suchas a slick paper, protects the SACNT film from sticking to the roller, and can be peeled off when using the film.Figure 9b shows a reel of SACNT/polyethylene (PE) compositefilmthat is produced from anentire wafer of SACNT array. The width of the film in this reel is about 8 cm, and the length can be over 60 m. In principle, by periodically inserting a new SACNT source wafer, the composite film can be produced continuously by the roll-to-roll process. Unfortunately, the performance of such as-drawn films is far below our expectation. In order to improve their performance, the SACNTarrays were trimmed by the oxygen plasma to reduce their height, since lower arrays give rise to films without large bundles. Besides, the SACNT films were trimmed by lasers to burn the outmost CNTs of the bundles and to make the bundles thinner. After treatment, films with excellent performance (24 Ω/sq @ 83.4%, 208 Ω/sq @ 90%) were obtained, and successfully used as touch panels.

Compared with dry method, solution-based method is much easier to prepare CNT films with high reproducibility. Perhaps the simplest way to make CNT films is by filtering the solution of dispersed tubes over a porous filter membrane. Filtration leads to highly uniform and reproducible films, and has precisely control over density [64]. Therefore, this method is often used to evaluate CNT materials and dispersion quality. Deposition method does not have the issues on the wetting on various substrates and it works well with extremely dilute CNT solutions. Another merit deserve to be addressed is that some excess dispersants could be washed away during the filtering process, which could enhance the conductance of the films. To our experience, films prepared with filtration method always show higher conductance than films prepared with spray coating or rod-coating method, since all of the dispersants resided in the films in the later methods. Since the films are deposited onto filters, a transfer from filters to other substrates is generally needed. Accordingly, transfer methods such as PDMS method [65]. Laser transfer method and microwave assisted method were developed [66]. The limitation of this method is that the size of the films is constrained by the filter, and is difficult to scale up. It is likely that this method will continue to be restricted to academic research.

In addition to vacuum assisted filtration, there are other deposition techniques that are useful for small scalelab testing. These include spray coating, [11] spin coating, [67] dipcoating, [68] and draw-downs using a Mayer rod or Slot Die [69]. Spray coating is a simple and quick method to deposit CNT films. Typically, CNT ink is sprayed onto a heated substrate. The substrate is heated to facilitate the drying of the liquid. The set temperature for the substrate is adjusted by the choice ofsolvent. By using diluted solution and multiple spray coating steps, homogeneous films can be obtained. Bundling mayhappen during the drying process after the sprayed mist of CNT has hit the PET substrate. Thus, it is difficult to get good film uniformity. The most widespread deposition method involves depositing solution on a substrate by Mayer Rod or Slot Die, followed by controlled drying. Aheating bar is used to control the drying process.This technique can be used to coat directly onto polyethylene terephthalate (PET), glass, and other substrates at room temperature and in a scalable way. Inkjet printing is an old and popular technology due to its ability to print fine and easily controllable patterns, noncontact injection, solution saving, and high repeatability [62]. It is very prevalent inprinted electronics. In a typical ink jet printing process, the droplet size is around~10 pL and, on the substrate, has a diameter of around 20-50 μm. Printing on paper is much easier than printing on a plastic or glass substrate, due to the high liquid absorption of the paper, which avoids the dewetting of the liquid on substrates. The liquid droplet and substrate interactionis crucial for uniform drying of the liquid. The most useful deposition technique is roll to roll coating of CNT inks onto continuous rolls of plastics. This technique can coat film up to 2 m wide at speeds up to 500 m/min.One such roll-to-roll coating line running continuously would have the equivalent output of 30 traditional sputter coaters, and could produce enough film to satisfy half of the available touch panel market. Examples of various film fabrication methods were shown in Figure 10.

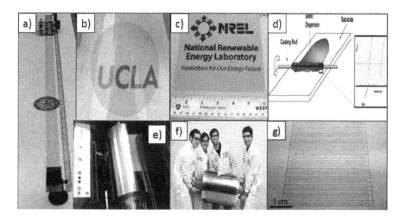

Figure 10. a) Transparent CNT film pulled from vertically grown CNT forest; b) CNT film transferred to PET using PDMS stamp. c) CNT film spray coated onto large areaplastic; d) Mayer rodcoating schematic. e) Image of CNT film being coated by slot die f) Roll of printed CNTfilm. g) Inkjet printed CNT lines. Reprinted with permissions from Ref. [4] copyright Wiley

6. Post-Treatment of CNT Films

During the preparation of CNT water solutions, dispersants are always introduced to assistant the exfoliation of CNT bundles. Since these dispersants are insulating, their residue decrease the conductance of CNT films significantly. Hence, post-treatments to remove these dispersants are necessary for preparing TCFs with high performance. In addition to remove the dispersants, doping is the other goal of post-treatment. In addition to rinsing with water, acid treatment is the most commonly used method to post-treat CNT films. As reported by Geng, [11] the sheet resistance of CNT films reduced by a factor of 2.5 times after treatment in concentrated nitric acid owing to the removal of surfactants SDS. Except their function on removing dispersants, concentrated nitric acid is often used to p-dope CNTs and enhances their conductivity [70]. Although nitric acid was effective to remove dispersants, they induced p-doping of CNTs, which will lead to instability of the films [71]. Besides, PET substrates will turn brittle after long time acid treatment. To solve this problem, Dr Sun's group developed a novel technique combing base treatment and short time acid treatment [72]. In their work, biomolecule RNA was chosen was the dispersant since they are easily degraded by base, acid and RNase. After depositing CNT films onto a PET substrate, they were immersed in the 5 wt % NaOH solution for one hour, and then treated with nitric acid for 10 min. The sheet resistance decreased significantly after treatment with NaOH solution owing to the removal of RNA molecules. After treatment with nitric acid, the RNA molecules were removed further and SWCNTs were slightly doped, therefore, the sheet resistance was reduced further. Base treatment combining short time acid treatment could remove RNA molecules efficiently as well as retaining the flexibility of PET substrates and the stability of the films.

7. Application of CNTTCFs

CNTTCFs have found a range of applications, among which we focus on the touch screens, plat panel displays, solar cells and OLEDs.

Touch screen is almost omnipresent in our daily life, such as in cell phones, tablet computers and many other electronics. Transparent electrodes are an essential component in most types of touch screens. High optical transmittance ($>$ 85%) and low sheet resistance Rs ($<$ 500 Ω/sq) are normally needed for touch screens. Meanwhile, extremely excellent durability, flexibility, and mechanical robustness are required given that the touch screen may be under indentation for millions of times. The mechanical robustness demonstrated by CNT touch panels give promises for increasing the lifetime and durability of current touch screens. There are a variety of touchscreen technologies that sense touch in different ways.Figure 11a shows the basic device structure and the transparent conductor arrangement for a 4-wire analog resistive touchpanel. These panels use two continuous electrodes separated by hemispheres of polymeric "spacer dots" that are10–100 μm in radius and 1–2.5 mm apart. Only at the edges (where electrode attachment occurs) is the transparent electrode patterned. Surface capacitive devices share the same type of continuous conductor whereas the projected capacitive deviceuses transparent conductors with specific patterning into predefined geometries. Resistive touch panels function by current driven measurements andcapacitive devices depend on capacitive coupling with the input device. Both panel types utilize signal processing controllers todetermine X-Y and sometimes Z position of inputs.

The mechanical durability of the transparent conductors is very important for resistive touch panels, since it involves compressive, sheer, and tensile stress every time it works. Their working process can be summarized as [4]:

1. Deformation of the touchside electrode–compressive, tensile

2. Contact of the touch sideand device side–compressive, shear

3. Contact of touch sideelectrode with spacer dots–compressive, shear

4. Extreme deformationof touch side electrode near edge seal–high tensile.

Compressive stress is not required to activatethe projected capacitive (ProCap) touch panels (of which theiPhone is a prime example). The ProCap touch panels are activated by a capacitive coupling with a suitable input device. Thus, there willnot be the mechanical flexing issues in ProCap devices. Still, the mechanical properties of the conducting layer are important since the conductors may be patterned to a size assmall as 10 μm in width. Metal oxides patterned to such small dimension become susceptible to cracking, fractures,and thermal cycling stress.

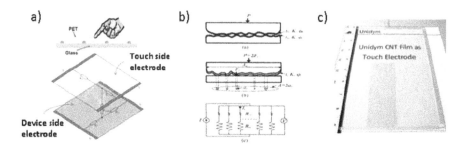

Figure 11. a) Schematic of four-wire resistive touch panel operation and functional layers; b) Schematic of the contact resistance experienced at the interface between two rough conductive layers separated by a very thin dielectric; c) Photograph of touch panel utilizing CNT film as touch electrode. Reprinted with permission form Ref. [4] copyright Wiley

Display panels are produced at nearly 1.7 billion unitsannually (1.2 billion mobile phones, 200 million televisions,150 million laptops, and 200 million desktop, machine interfaces, monitors etc. There are four common types of displays, which are electrowetting displays (EWD), electrochromic displays (ECD), electrophoretic displays (EPD) and liquid crystal displays (LCD). Currently, LCD devicesare manufactured in the greatest number and will be the mainsubject of this section. A transparent conductor'smajor role in LCD/EPD devices is to serve as pixel and common electrodes. An interesting advantage of using CNT films for LCD is the ability to use them possibly as both the transparent electrode and the alignment layer [73]. Recently, Lee et al demonstrated high performance TN-LC cells with ultra-thin and solution-processible SWNT/PS-*b*-PPP nanocomposite alignmentlayers. At an optimized SWNT density, a nanocompositegave rise to low power operation with a super-fast LC responsetime of 3.8 ms, which is more than four times faster than thaton a commercial polyimide layer due to the locally enhancedelectric field around individually networked SWNTs. Furthermore,TN-LC cells with their SWNT nanocomposite layers exhibited high thermal stability up to 200 °C without capacitance hysteresis.

Transparent electrodes are the essential components forphotovoltaic devices. The traditional electrodes for photovoltaic devices is ITO, which has high transmittance and low sheet resistance (~10-20 Ω/aq with the transmittance of 90%). However, their application was constrained by the high price of indium. Besides, the brittleness of ITO limited their usage in flexible devices, which will be a developing trend in the future. Therefore, replacing materials need to be developed. Carbon nanotubes are promising candidates since they have extremely high conductivity, high work function of 4.7-5.2 eV, relatively low cost and excellent flexibility. Besides, they are easy to be deposited into film via solution based process. Glatkowskiet al. [74] reported on the application of transparent CNT electrodesand found a PE-

DOT:PSS coating dramatically improves the device efficiency from 0.47% to 1.5%. The thin layer of PEDOT:PSS can smooth the CNT surface and enhance the charge transfer according to their investigation. In Hu's work, [75] flexible transparent electrodes were fabricated by printing SWCNT solutions on plastic substrates. The SWCNT films have a sheet resistance of 200 Ω/sq with a transmittance of 85%. The achieved efficiency of 2.5% (AM1.5G) approaches that of the controldevice made with ITO/glass (3%). Furthermore, the flexibility is far superior to devices using ITO coated on the same flexible substrate material. However, there are several aspects that need to be solved for CNT based electrodes.

1. Long termelectrical stability;

2. Occasional shorting betweenthe cathode and anode due to protruding CNTs;

3. Relatively high sheet resistance.

Light emitting diodes have an opposite light electricity coupling process as solar cells. Applications of nanoscale materials based transparent electrodes are mainly focused on organic light emitting diodes which hold great promise for the future electronics. In Aguirre's work, carbon nanotube anodes were implemented in small molecule OLED devices and achieved performance comparable to ITO-based anodes [76]. Recently, Feng et al [77] proposed a single walled carbon nanotubes-based anodes for organic light-emitting diodes (OLEDs) by spray-coating process without any use of surfactant or acid treatment. A layer of DMSO doped PEDOT:PSS was spray-coated on the SWCNT sheets to not only lessen the surface roughness to an acceptable level, but also improve the conductivity by more than three orders of magnitude. For the produced SWCNT-based OLEDs, a maximum luminance 4224 cd/m^2 and current efficiency 3.12 cd/A were achieved, which is close to the efficiency of ITO-based OLEDs.

Author details

Jing Sun* and Ranran Wang

*Address all correspondence to: Jingsun@mail.sic.ac.cn

State Key Lab of High Performance Ceramics and Superfine Microstructure, Shanghai Institute of Ceramics, Chinese Academy of Sciences, China

References

[1] Niu, C. M. (2011). *MRS Bull.*, 36, 766.

[2] Tyler, T. P., Brock, R. E., Karmel, H. J., Marks, T. J., & Hersam, M. C. (2011). *Adv. Energy Mater.*, 1, 785.

[3] Kauffman, D. R., Sorescu, D. C., Schofield, D. P., Allen, B. L., Jordan, K. D., & Star, A. (2010). *Nano Lett.*, 10, 958.

[4] Hecht, D. S., Hu, L. B., & Irvin, G. (2011). *Adv. Mater*, 23, 1482.

[5] De , S., & Coleman, J. N. (2011). *MRS Bull*, 36, 774.

[6] Tolcin, A. (2009). *Minerals Yearbook*, 35.

[7] Green, M. A. (2009). *Progress in Photovoltaics : Research and Apllications*, 17, 347.

[8] Hu, L., Hecht, D. S., & Gruner, G. (2004). *Nano Lett.*, 4, 2513.

[9] Geng, H. Z., Lee, D. S., Kim, K. K., Kim, S. J., Bae, J. J., & Lee, Y. H. (2008). *Journal of the Korean Physical Society*, 53, 979.

[10] De , S., Lyons, P. E., Sorel, S., Doherty, E. M., King, P. J., Blau, W. J., Nirmalraj, P. N., Boland, J. J., Scardaci, V., Joimel, J., & Coleman, J. N. (2009). *Acs Nano*, 3, 714.

[11] Geng, H. Z., Kim, K. K., So, K. P., Lee, Y. S., Chang, Y., & Lee, Y. H. (2007). *J. Am. Chem. Soc.*, 129, 7758.

[12] Scardaci, V., Coull, R., & Coleman, J. N. (2010). *Appl. Phys. Lett.*, 97.

[13] De , S., King, P. J., Lyons, P. E., Khan, U., & Coleman, J. N. (2010). *Acs Nano*, 4, 7064.

[14] Jackson, R., Domercq, B., Jain, R., Kippelen, B., & Graham, S. (2008). *Adv. Funct. Mater.*, 18, 2548.

[15] Zheng, Q. B., Gudarzi, M. M., Wang, S. J., Geng, Y., Li, Z. G., & Kim, J. K. (2011). *Carbon*, 49, 2905.

[16] Kim, K. K., Reina, A., Shi, Y. M., Park, H., Li, L. J., Lee, Y. H., & Kong, J. (2011). *Nanotechnology*, 21.

[17] Tantang, H, Ong, J. Y., Loh, C. L, Dong, X. C, Chen, P, Chen, Y, Hu, X, Tan, L P, & Li, L. J. (2009). *Carbon*, 47, 1867.

[18] Saran, N., Parikh, K., Suh, D. S., Munoz, E., Kolla, H., & Manohar, S. K. (2004). *J. Am. Chem. Soc.*, 126, 4462.

[19] Ko, W. Y, Su, J. W, Guo, C. H, & Lin, K. J. (2012). *Carbon*, 50, 2244.

[20] Geng, H. Z., Kim, K. K., Lee, K., Kim, G. Y., Choi, H. K., Lee, D. S., An, K. H., Lee, Y. H., Chang, Y., Lee, Y. S., Kim, B., & Lee, Y. J. (2007). *Nano*, 2, 157.

[21] Li, Z. R., Kandel, H. R., Dervishi, E., Saini, V., Biris, A. S., Biris, A. R., & Lupu, D. (2007). *Appl. Phys. Lett.*, 91.

[22] Zhang, D. H., Ryu, K, Liu, X. L, Polikarpov, E, Ly, J, Tompson, M. E, & Zhou, C. W. (2006). Nano Lett. , 6, 1880.

[23] Lyons, P. E., De , S., Blighe, F., Nicolosi, V., Pereira, L. F. C., Ferreira, M. S., & Coleman, J. N. (2008). *J Phys.*, 104.

[24] Maeda, Y., Hashimoto, M., Kaneko, S., Kanda, M., Hasegawa, T., Tsuchiya, T., Akasaka, T., Naitoh, Y., Shimizu, T., Tokumoto, H., & Nagase, S. (2008). *J. Mater. Chem.*, 18, 4189.

[25] Wang, W., Fernando, K. A. S., Lin, Y., Meziani, M. J., Veca, L. M., Cao, L., Zhang, P., Kimani, M. M., & Sun, Y. P. J. (2008). *Am. Chem. Soc.*, 130, 1415.

[26] Jackson, R. K., Munro, A., Nebesny, K., Armstrong, N., & Graham, S. (2010). *Acs Nano*, 4, 1377.

[27] Sorel, S., Lyons, P. E., De , S., Dickerson, J. C., & Coleman, J. N. (2012). *Nanotechnology*, 23.

[28] Park, J. G, Cheng, Q. F, Lu, J, Bao, J. W, Li, S, Tian, Y, Liang, Z. Y, Zhang, C, & Wang, B. (2012). *Carbon*, 50, 2083.

[29] Ebbesen, T. W., Ajayan, P. M., Hiura, H., & Tanigaki, K. (1994). *Nature*, 367, 519.

[30] Zimmerman, J. L., Bradley, R. K., Huffman, C. B., Hauge, R. H., & Margrave, J. L. (2000). *Chem. Mater.*, 12, 1361.

[31] Zhang, J., Gao, L., Sun, J., Liu, Y. Q., Wang, Y., Wang, J. P., Kajiura, H., Li, Y. M., & Noda, K. (2008). *J. Phys. Chem. C*, 112, 16370.

[32] Liu, Y. Q, Gao, L, Sun, J, Zheng, S, Jiang, L. Q, Wang, Y, Kajiura, H, Li, Y. M, & Noda, K. (2007). *Carbon*, 45, 1972.

[33] Wang, Y., Gao, L., Sun, J., Liu, Y. Q., Zheng, S., Kajiura, H., Li, Y. M., & Noda, K. (2006). *Chem. Phys. Lett*, 432, 205.

[34] Wang, R. R., Sun, J., Gao, L. A., & Zhang, J. (2010). *J. Mater. Chem.*, 20, 6903.

[35] Moon, J. M., An, K. H., Lee, Y. H., Park, Y. S., Bae, D. J., & Park, G. S. (2001). *J. Phys. Chem. B*, 105, 5677.

[36] Datsyuk, V., Kalyva, M., Papagelis, K., Parthenios, J., Tasis, D., Siokou, A., Kallitsis, I., & Galiotis, C. (2008). *Carbon*, 46, 833.

[37] Hecht, D., Hu, L. B., & Gruner, G. (2006). *Appl. Phys. Lett.*, 89.

[38] Bahr, J. L., Mickelson, E. T., Bronikowski, M. J., Smalley, R. E., & Tour, J. M. (2001). *Chem. Commun*, 193.

[39] Cheng, Q. H., Debnath, S., Gregan, E., & Byrne, H. J. (2010). *J. Phys. Chem. C*, 114, 8821.

[40] Moore, V. C., Strano, M. S., Haroz, E. H., Hauge, R. H., Smalley, R. E., Schmidt, J., & Talmon, Y. (2003). *Nano Lett.*, 3, 1379.

[41] Chen, J., Rao, A. M., Lyuksyutov, S., Itkis, M. E., Hamon, M. A., Hu, H., Cohn, R. W., Eklund, P. C., Colbert, D. T., Smalley, R. E., & Haddon, R. C. (2001). *J. Phys. Chem. B*, 105, 2525.

[42] Bergin, S. D, Sun, Z. Y, Rickard, D, Streich, P. V, Hamilton, J. P, & Coleman, J. N. (2009). *Acs Nano*, 3, 2340.

[43] Torrens, F. (2005). *Nanotechnology*, 16, S181.

[44] Landi, B. J., Ruf, H. J., Evans, C. M., Cress, C. D., & Raffaelle, R. P. (2005). *J. Phys. Chem. B*, 109, 9952.

[45] Kamlet, M. J., & Taft, R. W. (1976). *J. Am. Chem. Soc.*, 98, 337.

[46] Taft, R. W., & Kamlet, M. J. (1976). *J. Am. Chem. Soc.*, 98, 2886.

[47] Giordani, S., Bergin, S. D., Nicolosi, V., Lebedkin, S., Kappes, M. M., Blau, W. J., & Coleman, J. N. (2006). *J. Phys. Chem. B*, 110, 15708.

[48] Landi, B. J., Ruf, H. J., Worman, J. J., & Raffaelle, R. P. (2004). *J. Phys. Chem. B*, 108, 17089.

[49] Islam, M. F., Rojas, E., Bergey, D. M., Johnson, A. T., & Yodh, A. G. (2003). *Nano Lett.*, 3, 269.

[50] Ishibashi, A., & Nakashima, N. (2006). *Chem.-Eur. J.*, 12, 7595.

[51] Zheng, M., Jagota, A., Strano, M. S., Santos, A. P., Barone, P., Chou, S. G., Diner, B. A., Dresselhaus, M. S., Mc Lean, R. S., Onoa, G. B., Samsonidze, G. G., Semke, E. D., Usrey, M., & Walls, D. J. (2003). *Science*, 302, 1545.

[52] Zheng, M., Jagota, A., Semke, E. D., Diner, B. A., Mc Lean, R. S., Lustig, S. R., Richardson, R. E., & Tassi, N. G. (2003). *Nature Materials*, 2, 338.

[53] Cathcart, H., Quinn, S., Nicolosi, V., Kelly, J. M., Blau, W. J., & Coleman, J. N. (2007). *Journal of Physical Chemistry C*, 111, 66.

[54] Jeynes, J. C. G., Mendoza, E., Chow, D. C. S., Watts, P. C. R., Mc Fadden, J., & Silva, S. R. P. (2006). *Adv. Mater.*, 18, 1598.

[55] Wang, H., Lewis, J. P., & Sankey, O. F. (2004). *Phys. Rev. Lett.*, 93.

[56] Wang, H. M., & Ceulemans, A. (2009). Phys. Rev. B, , 79.

[57] Gowtham, S., Scheicher, R. H., Pandey, R., Karna, S. P., & Ahuja, R. (2008). Nanotechnology, , 19.

[58] Paul, S, & Kim, D. W. (2009). *Carbon*, 47, 2436.

[59] Marques, R. R. N., Machado, B. F., Faria, J. L., & Silva, A. M. T. (2010). *Carbon*, 48, 1515.

[60] Hopkins, S. D., Putterman, S. J., Kappus, B. A., Suslick, K. S., & Camara, C. G. (2005). *Phys. Rev. Lett.*, 95.

[61] Liu, P., Sun, Q., Zhu, F., Liu, K., Jiang, K., Liu, L., Li, Q., & Fan, S. (2008). *Nano Lett.*, 8, 647.

[62] Hu, L. B., Hecht, D. S., & Gruner, G. (2010). *Chem. Rev.*, 110, 5790.

[63] Feng, C., Liu, K., Wu, J. S., Liu, L., Cheng, J. S., Zhang, Y. Y., Sun, Y. H., Li, Q. Q., Fan, S. S., & Jiang, K. L. (2010). *Adv. Funct. Mater.*, 20, 885.

[64] Wu, Z. C., Chen, Z. H., Du, X., Logan, J. M., Sippel, J., Nikolou, M., Kamaras, K., Reynolds, J. R., Tanner, D. B., Hebard, A. F., & Rinzler, A. G. (2004). *Science*, 305, 1273.

[65] Cao, Q., Hur, S. H., Zhu, Z. T., Sun, Y. G., Wang, C. J., Meitl, M. A., Shim, M., & Rogers, J. A. (2006). *Adv. Mater.*, 18, 304.

[66] Allen, A. C., Sunden, E., Cannon, A., Graham, S., & King, W. (2006). *Appl. Phys. Lett*, 88.

[67] Manivannan, S., Ryu, J. H., Lim, H. E., Nakamoto, M., Jang, J., & Park, K. C. (2010). *Journal of Materials Science-Materials in Electronics*, 21, 72.

[68] Song, Y. I., Yang, C. M., Kim, D. Y., Kanoh, H., & Kaneko, K. (2008). *J. Colloid Interface Sci.*, 318, 365.

[69] Tenent, R. C., Barnes, T. M., Bergeson, J. D., Ferguson, A. J., To, B., Gedvilas, L. M., Heben, M. J., & Blackburn, J. L. (2009). Adv. Mater., , 21, 3210.

[70] Zhao, Y. L., & Li, W. Z. (2010). *Microelectron. Eng.*, 87, 576.

[71] Graupner, R., Abraham, J., Vencelova, A., Seyller, T., Hennrich, F., Kappes, M. M., Hirsch, A., & Ley, L. (2003). *PCCP*, 5, 5472.

[72] Wang, R. R., Sun, J., Gao, L. A., & Zhang, J. (2010). *Acs Nano*, 4, 4890.

[73] Fu, W. Q, Liu, L, Jiang, K. L, Li, Q. Q, & Fan, S. S. (2010). *Carbon*, 48, 1876.

[74] Van de Lagemaat, J., Barnes, T. M., Rumbles, G., Shaheen, S. E., Coutts, T. J., Weeks, C., Levitsky, I., Peltola, J., & Glatkowski, P. (2006). *Appl. Phys. Lett.*, 88.

[75] Rowell, M. W., Topinka, M. A., Mc Gehee, M. D., Prall, H. J., Dennler, G., Sariciftci, N. S., Hu, L. B., & Gruner, G. (2006). *Appl. Phys. Lett*, 88.

[76] Aguirre, C. M., Auvray, S., Pigeon, S., Izquierdo, R., Desjardins, P., & Martel, R. (2006). *Appl. Phys. Lett*, 88.

[77] Xue, F., Zhuo, W. Q., Yanc, L., Xue, H., Xiong, L. H., & Lia, J. H. (2012). *Org. Electron.*, 13, 302.

Chapter 2

Carbon Nanotubes for Use in Medicine: Potentials and Limitations

Wei Shao, Paul Arghya, Mai Yiyong,
Laetitia Rodes and Satya Prakash

Additional information is available at the end of the chapter

1. Introduction

Structurally, Carbon nanotubes (CNTs) can be viewed as wrapped from graphene sheets. Single-walled carbon nanotubes (SWNTs) have one layer of graphene sheet, whereas, the multiwalled carbon nanotubes (MWNTs) contain multi layers of graphene sheets. The well-ordered molecular structure brings CNTs many remarkable physical properties, such as, excellent mechanic strength, ultrahigh surface area, high aspect ratio, distinct optical properties [1], and excellent electrical conductivity [2]. In last decade, CNTs are intensively explored for in-vitro and in-vivo delivery of therapeutics, which was inspired by an important finding that CNTs can penetrate cells by themselves without apparent cytotoxic effect to the cells [3]. The high aspect ratio makes CNTs outstanding from other types of round nanoparticles in that the needle-like CNTs allow loading large quantities of payloads along the longitude of tubes without affecting their cell penetration capability. With the adequate loading capacity, the CNTs can carry multifunctional therapeutics, including drugs, genes and targeting molecules, into one cell to exert multi-valence effects. In the other side, owing to the ultrahigh surface area along with the strong mechanical properties and electrically conductive nature, CNTs are excellent material for nanoscaffolds and three dimensional nanocomposites. In recent year, CNT-based devices have been successfully utilized in tissue engineering and stem cell based therapeutic applications, including myocardial therapy, bone formation, muscle and neuronal regeneration. Furthermore, owing to the distinct optical properties of CNTs, such as, high absorption in the near-infrared (NIR) range, photoluminescence, and strong Raman shift [4], CNTs are excellent agents for biology detection and imaging. Combined with high surface area of CNTs for attaching molecular recognition molecules, CNT-based, targeted nanodevices have been developed for selective imaging and sensing. There are many areas where CNTs are extremely useful. Given the scope in this chapter, We describe strategies for preparation of CNTs for their use in medicine. Specifically, we focus and highlight the important biomedical applications of CNTs in the field of drug delivery, gene delivery, stem cell therapy, thermal therapy, biological detection and imaging (figure 1). The methods for formulating CNT-based therapeutics to suit different routes of drug administration are also described. The limitations with emphasis on toxicity and over all future directions are discussed.

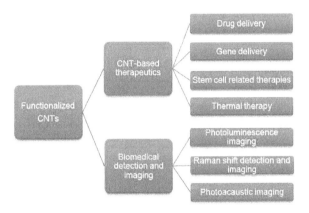

Figure 1. Functionalized CNTs in major biomedical applications.

2. Preparation of CNTs for use in medicine

Raw CNTs, persisting metallic nature, are highly hydrophobic. Therefore, surface modification of CNTs, or CNT functionalization, so as to disperse them into aqueous solutions becomes a key step for their biomedical applications. The CNT modification methods are involved in non-covalent and covalent strategies. The non-covalent modification utilizes the hydrophobic nature of CNTs, especially, π-π interactions for coating of amphiphilic molecules. The covalent modification generates chemical bonds on carbon atoms on CNT surface via chemical reactions followed by further conjugation of hydrophilic organic molecules or polymers rendering CNTs better solubility. These modifications not only offer CNTs water solubility, but also produce functional moieties that enable linking of therapeutic agents, such as genes, drugs, and recognition molecules for biomedical applications.

2.1. Non-covalent modification of carbon nanotube surface

The non-covalent modification approaches typically use amphiphilic molecules ranged from small molecules to polymers. The amphiphilic molecules associate with CNTs by either adsorbing onto or wrapping the CNTs [5]. The non-covalent modifications of CNTs are easy to perform. The process is only involved in sonication of CNTs with amphiphilic molecules in solvent at room temperature. Since it is a mild condition, CNTs molecular structure is not affected, and therefore their optical and electrical conductive properties are conserved.

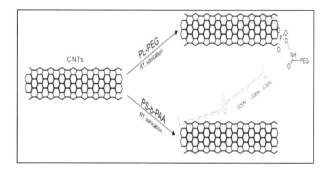

Figure 2. Schematic representation of adsorption of amphiphilic molecules onto carbon nanotube surface by π-π stacking and other hydrophobic interactions. Abbreviations: PL-PEG, phospholipid-polyethylene glycol; PS-b-PAA, polystyrene-block-polyacrylic acid.

Adsorption of amphiphilic molecules, such as surfactants, amphiphilic copolymers or others, onto CNT surfaces is one of the simplest and most effective way to disperse CNTs without destruction of their sp^2 hybridization [5]. The hydrophilic portions of surfactants interact with the polar solvent molecules, whereas, the hydrophobic portions adsorb onto the nanotube surface [5, 6]. The dispersity depends strongly on the length of the hydrophobic regions and the types of hydrophilic groups in the amphiphilic molecule. For example, surfactants with ionic hydrophilic head groups, such as sodiumdodecylsulfate (SDS) [7] or cetyltrimethyl ammonium bromide (CTAB) [8, 9], can stabilize a nanotube by electrostatic repulsion between micellar domains [7]. Nonionic surfactants, such as Triton X-100 [8], disperse CNTs mainly by forming a large solvation shell around a nanotube [8]. Figure 2 illustrates the manor of adsorbing amphiphilic molecules onto CNT surfaces, in which, hydrophobic alkyl chains or aromatic rings lay flat on graphitic tube surfaces. For example, an synthetic biocompatible lipid-polymer conjugate, phospholipid-polyethylene glycol (PL-PEG) has been applied for surface modification of CNTs, which gives rise to a variety of biomedical applications ranged from drug delivery, biomedical imaging, detection and biosensors [10]. The ionic surfactants, particularly those based on alkyl-substituted imidazolium cationic surfactants [11], can effectively disperse CNTs in organic or aqueous media by the counter anion [12, 13]. Polyaromatic derivatives carrying hydrophilic moieties can also effectively disperse CNTs in aqueous media by forming specific directional π–π stacking with the graphitic surfaces of nanotubes [6, 14]. In this context, pyrene, a polyaromatic molecule, demonstrated a high affinity toward CNT surfaces [6]. Interactions of the polyaromatic-moitie of pyrene with CNTs are strong enough to be irreversible, and therefore, the pyrene derivatives are used to anchor proteins or biomolecules on nanotube surface [6, 14, 15]. Other classes of polyaromatic molecules, such as substituted anthracenes, heterocyclic polyaromaticporphyrins [16] and phthalocyanines [17], disperse CNTs via the same mechanism.

The polymers containing hydrophobic backbone and hydrophilic side groups, eg. poly[p-{2,5-bis(3-propoxysulfonic acid sodium salt)}phenylene] ethynylene (PPES), can effectively disperse CNTs in water, in which, the strong $\pi-\pi$ interactions between CNTs and aromatic backbone of the polymers drive the wrapping of CNTs, and the water-soluble side groups impart solubility of CNTs in water [18]. DNA and siRNA can disperse CNTs by wrapping. DNA or siRNA are made of hydrophobic bases and alternative hydrophilic phosphates and riboses. Such structure facilitates CNT dispersion by the bases wrapping to CNTs and the hydrophilic sugar-phosphate groups extending to water phase [19].

2.2. Covalent modification of CNT surface

The covalent modification, namely the chemical modification of CNTs is an emerging area in materials science. Among the various strategies, the most common ones are:

i.   esterification and amidation of oxidized CNTs,

ii.   generation of functional groups on CNT sidewalls by cycloaddition reactions.

Oxidation of CNTs is a purification method for raw CNTs. Oxidation of CNTs is carried out by reflexing raw CNTs in strong acidic media, e.g. HNO_3/H_2SO_4. Under this condition, the end caps of the CNTs are opened, and carboxylic groups are formed at these ends caps and at some defect sites on nanotube sidewalls (Figure 3a) [20]. The carboxylic groups provide opportunities for further derivatization of the CNTs through esterification or amidation reactions. For example, some organic molecules with amine groups can be directly condensed with the carboxylic groups present on the surface of the CNTs [6, 14]. Alternatively, the carboxyl moieties can be activated with thionyl chloride and subsequent react with amine groups (Figure 3b) [6, 14]. These reactions are widely applied for conjugation of water-soluble organic molecules, hydrophilic polymers, nucleic acid (DNA or RNA), or peptides to the oxidized CNTs, which result in multifunctional CNTs [6, 14]. In most cases, the length of nanotubes is often shortened [20] ,but the electronic properties of such functionalized CNTs remain intact. Oxidation reaction only generates carboxyl groups on cap ends and defect sites on CNTs. To generate chemical bonds on sidewall and cap ends of CNTs, cycloaddition reactions are used [21](Figure 4). Cycloaddition reaction is a very powerful methodology, in which the 1,3-dipolar cycloaddition of azomethineylides can easily attach a large amount of pyrrolidine rings on sidewalls of nanotubes. Thus, the resulting functionalized CNTs are highly soluble in water [22]. In addition, pyrrolidine ring can be substituted with many functional groups for different applications.

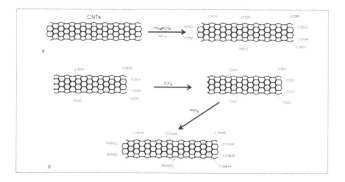

Figure 3. Covalent modification of carbon nanotubesby a) Oxidation reaction of cap end of CNTs and b) further attaching hydrophilic molecules by amidation reactions.

In contrast to non-covalent surface modifications, which do not locally disrupt sp^2 hybridization, or create defects, the covalent surface modifications disrupt CNT sp^2-conjugated structures and therefore, could affect the electronic and optical performances [5].

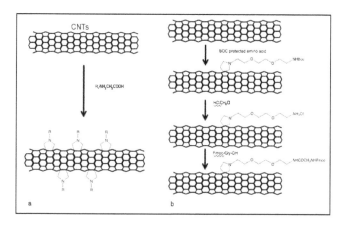

Figure 4. Sidewall covalent modification of carbon nanotubes a) a general scheme of 1,3-dipolar cycloaddition reaction b) preparation of amino-functionalized CNTs by 1,3-dipolar cycloaddition.

3. Carbon nanotube based therapeutics

3.1. Carbon nanotubes for chemotherapy drug delivery

Cancer is one of the most common causes of death worldwide. Chemotherapy in addition to the surgical removal of tumors is a conventional treatment for cancers. However, the effectiveness of chemotherapy drugs is often limited by the toxicity to other tissues in the body. This is because most chemotherapy drugs do not specifically kill cancer cells, they act to kill all cells undergoing fast division. Nanoparticles have been applied to drug delivery and showed improved drug efficiency and reduced off-target tissue toxicity due to accumulation in tumor tissues. Nanoparticles target tumor tissues by two mechanisms: passive targeting and active targeting. As fast growing tissues, tumors display enhanced vascular permeability due to high demand for nutrients and possible oxygen. The features of the leaky vasculature are employed for delivery of nanoparticle drugs since the size of nanoparticle allows them to accumulate in tumor tissues [23]. The phenomenon is termed as tumor-selective *enhanced permeability and retention* (EPR) effect. More efficient tumor targeting can be achieved through active targeting approaches, in which, targeting molecules can recognize tumor biomarkers on cancer cell surface. The properties of CNTs are beneficial for cancer drug delivery, firstly, like other nanoparticles, the size of functionalized CNTs is preferable for accumulation in tumor tissues; secondly, CNTs contain ultrahigh surface area of CNTs facilitate loading of drugs and targeting molecules; thirdly, the hydrophobic benzene ring structure of CNTs can be used for loading drugs that contain benzene ring structure, eg. doxorubicin (DOX), epirubicin (EPI), and daunorubicin (DAU).

Preparation of tumor-targeted devices using carbon nanotubes

A range of tumor targeting molecules has been discovered, including tumor specific antibodies, peptides, and others. Antibodies have been developed to specifically binding to biomarkers on cancer cell surface, eg, Trastuzumab recognizes Human Epidermal Growth Factor Receptor 2 (HER-2) positive cancer cells[24] and anti-CD20 for CD20 biomarker on B cell lymphoma [25]. These antibodies have therapeutic effects on their own, and can also serve as tumor targeting probes. Alpha V beta 3 ($\alpha v\beta 3$) integrin is a heterodimerictransmembrane glycoprotein found on a variety of tumor cells, including osteosarcomas, neuroblastomas, glioblastomas, melanomas, lung, breast, prostate cancers. $\alpha v\beta 3$) integrin is a wellrecognized target for cancers. The amino acid sequence of Arg-Gly-Asp (RGD) is identified to be responsible for tight binding to $\alpha v\beta 3$ integrin, which leads to the development of short tumor targeting peptide RGD [26]. Similar to RGD, another type of peptide contains Asn-Gly-Asp (NGR) triad that binds to the endothelium cells on neoangiogenic vessels. NGR-tagged delivery systems have been developed to deliver cytokines, nanoparticles, and imaging agents to tumor blood vessels [27]. Folic acid, a small molecule vitamin, binds to folate receptor overexpressed in a variety of cancer cells, including breast, colon, renal and lung tumors[28]. As described in section 2 of this chapter, a variety of chemical and physical methods have been developed for functionalization of CNTs. The above listed tumor targeting molecules are mostly proteins or peptides, which contains sulfhydryl groups that can be easily conjugated to amino-functionalized CNTs [14] using heterbifunctional linker molecules that contain NHS ester on one end and Maleimide on the other end [29]. These conjugation reactions are usually carried out under mild conditions [10]. Thus, the molecular structure of CNTs is not disturbed, and therefore, the optical properties are preserved. The CNT-based targeted devices developed by this methods are good for potential tumor detection and imaging applications. To date, all above-mentioned tumor-targeting strategies have been applied for construction of CNT-based, which will be discussed separate in the following sub-sections.

Targeted delivery of chemotherapy drugs by physically absorbed on carbon nanotubes

Supermolecular benzene ring structure of CNTs affords surprisingly high degree of aromatic molecules by π-π stacking. DOX, an important chemotherapeutic agent, has been efficiently loaded onto SWNTs-PL-PEG for tumor-targeted delivery [30, 31]. Binding to and release of drug molecules from nanotubes could be controlled by adjusting pH.The appropriate diameter of nanotube for drug loading was used because the strength of π-π stacking of aromatic molecules was dependent on nanotube diameter. In-vivo study with SWNTs-PL-PEG/DOX showed significantly enhanced therapeutic efficacy in a murine breast cancer model [30]. With further attaching tumor targeting molecules, eg. folic acid (FA) or RGD, the targeted SWNTs-DOX could more effectively inhibit the growth of cancer cells in-vitro and in-vivo [31-34]. Similar physical absorption method was applied for drug DAU using SWNTs, in which, sgc8c Aptamer was used to target leukemia biomarker protein tyrosine kinase-7 [33]. It has been shown that Aptamer-SWNTs-DAU was able to selectively target leukemia cells. The release of DAU was pH-dependent.Other hydrophobic drug molecules, such as paclitaxel (PTX), docetaxel (DTX), can also be absorbed on CNTs surface for delivery [32, 35], however, their loading efficiency and stability were much lower due to their comparatively bulky structure.

Targeted delivery of chemotherapy drugs by covalently linked to functionalized carbon nanotubes

Non-aromatic small molecule drugs can be chemically conjugated to CNTs for delivery. However, the drugmolecules have to be released from the CNTs to take effect, so the linkages between the drugs and CNTs have to be cleavable. Preferably, the active drugs are released inside of the target cells to reduce toxic effect to the neighbouring healthy cells. The common linkers that are used for drug delivery include ester, peptide, and disulfide bonds. These linkers can be cleaved by the 7enzymes present in the routes of delivery. Specifically designed linkers allow controlled release of drug into desired sites. For example, drug cisplatin has been conjugated directly to oxidized SWNTs via a peptide linker [36]. This specific peptide linkage has been shown tobe selectively cleaved by proteases overexpressed in tumor cells. Further conjugation of epidermal growth factor (EGF), a growth factor that selective binding to EGF receptor overexpressed on cancer cells, to SWNTs-cisplatin led to more efficient tumor inhibition compared to both free cisplatin and non-targeted SWNTs-cisplatin [36]. Alternative to conjugation of drugs to CNTs directly, drugs can also be conjugated to the molecules, eg. polymers, that are used to disperse CNTs. The end functional groups in the polymers are used for drug linkage. This method is very useful for delivery of bulky, hydrophobic drug molecules. In one example, SWNTs was dispersed using a biocompatible polymer PL-PEG-NH$_2$[10] and drug PTX was conjugated to SWNT-PL-PEG-NH$_2$ via via ester bonding for delivery [37]. PTX is one of the most important drugs for metastatic breast cancer. However, currently available formulations for PTX have to be infused intravenously over long periods of time due to the side effects. In addition, due to poor water solubility of the drug, necessity of use organic solvent, such as Cremophor in clinical formulation Taxol® causes sever side effects and hypersensitivity reactions [38, 39]. Conjugation of PTX to SWNTs-PL-PEG-NH$_2$ enable removing of solvent in delivery. Indeed, the SWNTs-PL-PEG-PTX displayed increased tumor inhibition effect and reduced side effects in a murine breast cancer model compared with Taxol® formulation [37].

Drugs	Targeting Moieties	Cancer Biomarkers	Type of CNTs	References
Cisplatin	EGF	EGF Receptor	SWNTs	[36]]
Daunorubicin	Sgc8c Aptamer	Tyrosine Kinase-7	SWNTs	[33]
Docetaxel	NGR	Endothelial Cells	SWNTs	[32]
Doxorubicin	Folate /Magnetic	Folate Receptor	MWNTs	[34]
Doxorubicin	RGD	Integrin αvβ3	SWNTs	[30]
Doxorubicin	Folate	Folate Receptor	SWNTs	[31, 40]
Gemcitabine	Magnetic	/	MWNTs	[41]
Platinum (IV)	Folate	Folate Receptor	SWNTs	[42]

Table 1. . Abbreviations: SWNTs, single walled carbon nanotubes; MWNTs, multiwalled carbon nanotubes, EGF, epidermal growth factor; RGD, peptide with arginine-glycine-aspartate sequence; NGR, peptide with asparagineglycine - arginine sequence; Sgc8c, oligonucleotide sequence.

3.2. Carbon nanotubes for gene delivery

Gene therapy is an important treatment for cancer and other genetic diseases. However, the effects of gene therapy are limited by the efficiencies of transfection and system delivery. Since DNA and siRNA are macromolecules, they cannot pass through cell membrane by themselves, carriers are needed to take them inside of cells to take effects. Structurally, both DNA and siRNA contain anionic phosphodiester backbone that and be complexed with cationic reagents, such as cationic lipids and polymers, etc. For system delivery, the DNA or siRNA can be loaded into cationic nanoparticles made from cationic lipids or polymers [43, 44]. The nanoparticles could protect them from nucleases degradation. Since CNTs are able to penetrate cells [3], they are investigated for gene delivery. Typically, two methods are used for loading nucleic acids to CNTs:

i. electrostatic association with cationic molecule functionalized CNTs [45, 46];

ii. chemical conjugation of nucleic acids to functionalized CNTs via cleavable chemical bonds [47];

iii. DNA or siRNA are directly wrap to raw or oxidized CNTs.

Gene delivery using cationic molecule functionalized carbon nanotubes via electrostatic interactions

As discussed early, cationic molecules, such as, ammonium-containing molecules and poly-ethylene imine (PEI), can be covalently linked to chemically modified CNTs by oxidation or 1,3-cycloadditions reactions[47-50]. In one application, DNA was loaded into CNTs conjugated with ammonium-terminated oligoethylene glycol(CNTs-OEG-NH$_3$$^+$) for delivery [47,

48]. Using this deliver vehicle, expression of test plasmid pCMV-βgal was examined in-vitro.Result showed that the transfection efficiency of CNTs carrier was 5-10 times higher than naked DNA; but, much lower than that of liposome [47]. It has been shown that charge ratio (ammonium groups on CNTs vs phosphate groups of the DNA backbone) is a determination factor for gene expression [48]. In contrast to DNA delivery, the same CNTs carrier for delivery of cyclin A2 siRNA demonstrated pronounced silencing effect in-vitro[51]. Surprisingly, In-vivo delivery of SOCS1 significantly inhibited SOCS1 expression and retarded the tumor growth in murine B16 tumor model [52]. The studies with PEI functionalized CNTs also showed very positive results. PEI is an efficient gene delivery reagent by its own, however, high amount of PEI is toxic to cells. The siRNA delivery by PEI-grafted MWNTs showed improved gene expression to the equivalent amounts of PEI polymer alone but with reduced cytotoxicity [46, 53].

Gene delivery by covalently conjugation to carbon nanotubes via cleavable chemical bonds

Alternatively, genes can be conjugated to amphiphilic polymers that are used for non-covalent CNT functionalization [10, 54,55]. Incorporation of cleavable chemical bonds facilitates releasing of DNA or siRNA cargos from CNTs in a controlled manner [54]. Thiol-modified DNA or siRNA were covalently conjugated to amino group of SWNT-PL-PEG- NH_2 via cleavable disulfide bond [55]. The genes were released by the cleavage of disulfide bonds by thiol digesting enzymes upon cellular internalization of CNT-PL-PEG-siRNA. The CNT-mediated siRNA delivery showed better gene transfection efficiency than liposome-based delivery system in hard-to-transfect human T cells and primary cells lines [54].

Gene delivery by wrapping directly on carbon nanotubes

Nucleic acids, DNA or siRNA, contains alternative amphiphilic motifs, which can be used to disperse CNTs in water. The nucleic acids form helical wrapping around the CNTs with the bases binding to the hydrophobic CNTs and the hydrophilic sugar-phosphate groups extending to the water phase [19]. In this way, DNA or siRNA serves both CNT dispersing agent and the cargo. It has been shown that the siRNA functionalized SWNTs readily enter cells and exerts its biological activity in cell culture [19]. Studies with intratumoralinjection of siRNA functionalized SWNTs showed significantly inhibition effect in-vivo [56].

3.3. Carbon nanotubes for stem cell related therapies

There has been an increasing trend in attempts to design and develop different CNT based tools and devices for tissue engineering and stem cell therapy applications. In particular, CNT impregnated nanoscaffolds have shown multiple advantages over currently available scaffolds. This includes its strong mechanical properties, resemblance of structure with collagen fibrils and extracellular matrix and electrically conductive nature. These attributes of the CNT based scaffolds and three dimensional nanocomposites have led to their diverse therapeutic applications in the field of myocardial therapy, bone formation, muscle and neuronal regeneration. These applications are mainly based on one principle and that is to modulate the stem cell growth and differentiation in a more controlled and desirable manner.

Carbon nanotubes for stem cell based heart therapy

Over the past two decades there has been significant advancement in stem cell therapy to repair and replace damaged tissues, such as heart muscle[57]. This is because of their ability to divide and differentiate into diverse specialized cell types. Recently, there has been increasing body of evidence indicating that the extracellular matrix plays a critical role in stem cell viability, proliferation and differentiation [58, 59]. Hence, designing a microenvironment prepared from polymeric scaffolds which imitate the physical characteristics of natural biomatrix has been the central strategy in tissue engineering. The emergence of nanomaterials such as nanotubes provide opportunities to design such biocompatible scaffolds for hosting and directing stem cell differentiation [60].

Preliminary studies demonstrate that neonatal rat ventricular myocytes cultured on substrates of multiwall carbon nanotubes can interact with the nanofibres by forming tight contacts and show significantly improved mitotic and chemotactic effects [61]. Moreover, such mode of culture also altered the electrophysiological properties of cardiomyocytes, indicating that CNts are able to promote cardiomyocyte maturation. Further investigations with a nanocomposite of PLGA:CNF show that cardiomyocyte density increases with greater amounts of CNF in PLGA [62]. The study also showed similar trends with neurons. The immense potential of this technology for myocardial therapy roots from the fact that this cardiac patch can not only promote myocardial cells, but also induce the nerve cell growth that help the cardiac cells to contract. In addition, it also supports endothelial cells that make the inner lining of the blood vessels supplying oxygen to the heart.

Carbon nanotubes for stem cell based bone regeneration

In order to direct stem cell differentiation towards bone regeneration, there has been increasing interest by the researchers to explore topographical features of the cell culture substrate. Physical factors, such as rigidity of the extracellular environment, can influence stem cell growth and differentiation. Such differentiation of human stem cells can be detected by altering the size of the nanotubes on which the cells are grown [63]. It has been reported that 70- to 100-nm diameter nanotubes can initiate rapid stem cell elongations, which induce cytoskeletal stress and selective differentiation into osteoblast-like cells, offering a promising route for quicker and better recovery, for example, for patients who undergo orthopedic surgery. The group also showed that the differentiated stem cells express osteopontin and osteocalcin, the two important osteogenetic protein markers.

Moreover CNTs are promising materials for nanaoscaffold and implantation purposes due to the fact that CNTs are conductive, have excellent mechanical properties and their nanostructured dimensions mimic the 3D structure of proteins found in extracellular matrices. Their dimensions resembles closely with that of the triple helix of collagen fibrils which can promote for nucleation and growth of hydroxyapatite, the major inorganic component of bone. A newly developed nanocomposite scaffold of CNFs/CNTs has been shown to influence the cell behaviour [64]. In-vitro study demonstrated that, smaller dimension CNFs dispersed in polycarbonate urethane promoted osteoblast adhesion but did not promote the adhesion of fibroblasts, chondrocytes, and smooth muscle cells. But the mechanisms that guide such cell functions are yet to be understood.

Surface functionalizing the nanotube surface with bone morphogenetic protein-2 (BMP-2) further accelerates chondrogenic and osteogenic differentiation of MSCs [65, 66]. This stimulation is a combined effect of the surface nanoscale geometry of the substrate nanostructures and their BMP-2 coating efficiency. In a similar kind of study, the system also exhibited higher cell proliferation rate, apart from enhanced differentiation [66]. Nanotubes can also be used for extended drug release as has been demonstrated by Hu et al, where drug loaded nanotubes, in combination with multilayers of gelatin and chitosan, have been shown as a new way to use nanotubes as reservoir for storing drugs [67]. The system effectively promoted osteoblastic differentiation of MSCs. Further studies in this direction can be beneficial in order to develop potential bone implants for improved bone osteointegration.

Carbon nanotubes for stem cell based neuronal regeneration

The unique abilities of human embryonic stem cells (hESCs), such as their self-renewal and potency, hold great promise in the field of regenerative medicine and stem cell based therapy. The derivation of neuronal lineages from hESCs holds promise to treat neurological pathologies of the central and peripheral nervous system such as Parkinson's disease, spinal cord injury, multiple sclerosis and glaucoma [68, 69]. CNT based substrates have been shown to promote neuronal differentiation [70]. It has also been proposed that neurons grown on a CNT meshwork displayed better signal transmission, due to tight contacts between the CNTs and neural membranes conducible to electrical shortcuts [71]. It was demonstrated that the MSCs and the neurosphere of cortex-derived neural stem cells (NSCs) can grow on the CNT array and both MSCs and NSCs interacted with the aligned CNTs. The results suggest that CNTs assist in the proliferation of MSCs and aid differentiation of cortex-derived NSCs [72]. However, due to the harsh external environment in the host body and lack of supportive substrates during transplantation, much of the transplanted cells lose its viability resulting in reduced therapeutic efficacy [73]. It has been reported that two dimensional thin film scaffolds, composed of biocompatible poly(acrylic acid) polymer grafted carbon nanotubes (CNTs), can selectively differentiate human embryonic stem cells into neuron cells while maintaining the viability of transplanted cells [74]. Even multiwalled carbon nanotube (MWNT) sheets showed to significantly enhance neural differentiation of hMSCs grown on the CNT sheets. Axon outgrowth was also controlled using nanoscale patterning of CNTs [75]. Recently, silk-CNT-based nanocomposite scaffolds are shown to protect and promote neuronal differentiation of hESCs [76]. Silks are natural polymers (protein) that have been widely used as biomaterials for many years. Fibroin, comprising the major portion of the silk protein fibre, consists of 90% of amino acids including glycine, alanine, and serine. Due to its strong mechanical and flexible nature in thin film form, biocompatibility, and *in-vivo* bioresorbable properties, fibroin protein has been used as the building block for scaffolds. As confirmed by scanning electron microscope (Figure5 A-C), similar results were obtained with the developed silk-CNT scaffold where cells grown on the silk substrate exhibited denser complex three-dimensional axonal bundle networks as well as better spatial density distribution of the networks compared to other scaffolds. Overall, the silk-CNT nanocomposite provided an efficient three-dimensional supporting matrix for stem cell-derived neuronal transplants, offering a promising opportunity for nerve repair treatments for patients with neurological disorder. In-vitro analysis showed that β-

III tubulin, representing the mature differentiated neurons and nestin, representing the neuron precursors, were highly expressed in hESCs grown on the silk-CNT substrate compared to the expression level of cells grown on the control poly-L-ornithine substrate (figure 5D). In addition, hESCs cultured on the silk-CNT scaffold exhibited higher maturity along with dense axonal projections.

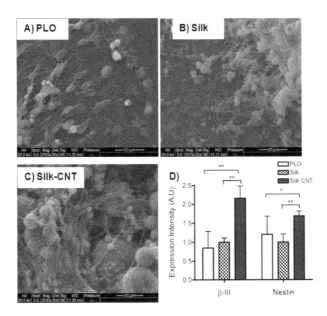

Figure 5. Scanning Electron Microscope images of hESCs on various substrates. SEM images of (A) cells cultured on PLO exhibiting a flat morphology and two-dimensional axonal connections, (B) cells cultured on silk scaffolds demonstrating three-dimensional structures and cell migration, and (C) cells cultured on silk-CNT scaffolds demonstrating three-dimensional axonal connections and silk-CNT matrix degradation. (D) Two neuronal markers (β-III tubulin and nestin) were used to further determine the hESC differentiation efficiency.Expression intensity of β-III tubulin and nestin was observed with fluorescence microscopy. Silk-CNT scaffolds exhibited maximum β-III tubulin expression, while nestin expression exhibited a similar trend. *=P< 0.01, and **= P< 0.001[76]. Abbreviations: PLO, Poly-L-ornithine; hESCs, human embryonic stem cells; CNT, carbon nanotube.

3.4. Carbon nanotubes for thermal destruction of tumors

Tissues are known to be highly transparent to 700- to 1,100-nm near-infrared (NIR) light, whereas, SWNTs display strong optical absorbance in this special spectral window. When constantly absorb energy in NIR region, SWNTs emit heat [77]. Continuous heating leads to killing of the cells. SWNTs have been engineered with tumor recognition molecules for selective enteringcancer cells. Upon NIR radiation, the cancer cells were killed by thermal ablation [78-83]. Previous studies have shown that folic acid decorated SWNTs more effectively killed folate receptor positive cancer cells [83]; monoclonal antibody (mAb) against human CD22 conjugated SWNTsonly targeted CD22(+)CD25(-) Daudi cells; whereas, anti-CD25 mAb coupled SWNTs only target CD22(-)CD25(+) activated peripheral blood mono-nuclear cells [81]. The thermal ablation effects can be combined with other therapies, eg. chemotherapy, by loading drugs on CNTs for synergic effect [32].

Tumors, in general, contain a small population of tumor initiating stem-like cells, termed as cancer stem cells. These cells are unmanageable by standard treatment modalities such as chemotherapy and radiotherapy and tend to persist after treatment [84]. Heat-based cancer treatments are increasingly becoming a potential alternative to approach this problem. Com-bining CNTs with such hyperthermia based therapies can further enhance its efficacy by si-multaneously eliminating both the stem cells and bulk cancer cells that constitute a tumor. In fact, CNTs offer several properties that make them promising candidates for such thermal therapy. This includes their ability for thermal conductance and strong absorbance of elec-tromagnetic radiation. It generates significant amounts of heat upon excitation with near-in-frared light which is transparent to biological systems including skins. Such a photothermal effect can be employed to induce thermal cell death in a noninvasive manner. Thus, if CNTs can be localized to tumors, they can be stimulated with near-infrared radiation or radiofre-quency energy to generate site-specific heat [85]. Preliminary in-vivo results show that a combination of multiwalled carbon nanotubes (MWNTs) and NIR can be useful for tumor regression and long-term survival in a mouse model [86]. Such CNT-mediated thermal ther-apy addresses the limitations of presently available medical strategies. This includes the minimally invasive site-specific heating which will greatly diminish the off-target toxicities, generation of uniform temperature distribution throughout the tumor mass by the activated CNTs, its compatibility with concurrent MRI temperature mapping techniques. It has also been recently reported that breast cancer stem cells, highly resistant to conventional thermal treatments, can be successfully treated with CNT-based photothermal therapies by promot-ing necrotic cell death [84]. Further studies in this direction shows that DNA-encased MWNTs are more efficient at converting NIR irradiation into heat compared to non-encased MWNTs and that this method can be effectively used in-vivo for the selective thermal abla-tion of cancer cells [87].

Glioblastomamultiforme is the most common and aggressive malignant primary brain tu-mor involving glial cells and accounting for a large percentage of brain and intracranial tu-mor [88, 89]. It is also known for its recurrence and overall resistance to therapy. CD133+ stem cells occurring among GBM cells are responsible for such huge recurrence risk [90]. Re-search has been focused on developing strategies to efficiently deliver CNTs to these target sites, harboring CD133+ cancer stem cells [80]. In-vitro studies show that such targeted elim-ination of CD133 (+) cancer stem cells are possible by adding SWNTs functionalized with CD133 monoclonal antibody, followed by irradiation with NIR laser light. In a separate study, embryonic stem cells, once administered with MWNTs, have shown to induce an en-hanced immune boost and provide subsequent anticancer protection in mice with colon can-cer by suppressing the proliferation and development of malignant colon tumors [91].

4. Carbon nanotubes for biomedical imaging and detection

The well-ordered molecular structure attributes CNTs with multiple distinct optical properties, include strong NIR absorption, photoluminescence and Raman shift [92]. Structurally, SWNTs can be viewed as a cylinder rolled up by one layer of graphene sheet. TThere are infinite numbers of ways to roll a graphene sheet into a cylinder. Depending on different ways of wrapping, the particular nanotube could be metallic or semi-conductive. Individual semi-conductive SWNTs with appropriate chirality can generate a small band gap fluorescence of 1 eV, which corresponds to NIR range (900-1600 nm), where biological tissues have very low absorption, scattering, and autofluorescence, and therefore, are very useful for biological imaging. In the other side, the inherent graphene structure provides SWNTs with specific Raman scattering signature [93], which is strong enough for use in-vivo imaging. All these optical properties offer opportunities for SWNTs as contrast agents for near-infrared (NIR) photoluminescence imaging [94, 95], Raman imaging and optical absorption agent for photoacoustic imaging [96-98].

4.1. Photoluminescence imaging

NIR photoluminescence of micelle encapsulated SWNTs was firstly discovered by O.Connel et al [7]. The single-particle dispersion of individual nanotubes was prepared by ultrasonically agitating of raw SWNTs in SDS. The tube bundles, ropes, and residual catalyst were removed by ultracentrifugation, since the aggregation of nanotubes would quench fluorescence. One advantage of the photoluminescence of SWNTs over organic fluorescence dyes is that SWNTs have no apparent photobleaching, and therefore, the SWNTs could be a powerful tool for tracking changes in living system. Researchers have applied NIR photoluminescence of SWNTs for tracking endocytosis and exocytosis of SWNTs in NIH-3T3 cells in real time [94, 95]. Moreover, conjugation of antibodies to SWNTs surface allowed specific cell targeting. They have shown that, with conjugation of anti-CD20, SWNTs selectively recognized CD20 cell surface receptor on B-cells with little binding to receptor negative T-cells. Similarly, with conjugation of Herceptin, SWNTs only recognize HER2/neu positive breast cancer cells. The selective binding of SWNTs was detected by intrinsic NIR photoluminescence of nanotubes. This technique allows deep tissue penetration and high-resolution intravital microscopy imaging of tumor vessels beneath thick skin [99, 100].

4.2. Raman shift imaging

Raman spectroscopy is a sensitive analytical tool for biological samples. It also has advantages of resistance to autofluorescence and photobleaching, high spatial resolution, and small sample size [101]. CNTs exhibit strong resonance Raman scattering with several distinctive scattering features including the radial breathing mode (RBM) and tangential mode (G-band) [93, 102]. Both RBM and G-band of CNTs are sharp and strong peaks, which can be easily distinguished from autofluorescence of tissue samples, Recently, Raman microspectroscopy of SWNTs has been applied for imaging of tissue samples, live cells, and small animal models [96-98]. Tumor targeted delivery by RGD peptide functionalized SWNTs has been investigated in murine tumor model [85]. Raman spectroscopy image of excised tissues confirmed efficient targeting of $\alpha v\beta 3$ integrin positive U87MG tumor by RGD [85]. This study also disclosed that CNTs have relatively long circulation time, and rapid renal clearance, which makes SWNTs an attractive diagnostic and therapeutic delivery vehicle. Zevaleta et al further developed a Raman microscope capable of noninvasive in-vivo evaluation of tumors in mice with RGD-labeled SWNTs. Using the dynamic Raman microscope, pharmacokinetics of SWNTs in the tumor was evaluated immediately following an intravenous injection of SWNTs. Raman spectral analysis revealed effectiveness of the RGD nanotubes to the integrin expressing U87MG tumor. The noninvasive Raman imaging results were compared with excised tissues and shows consistency [97].

4.3. Photoacaustic imaging

Photoacoustic imaging is an optical imaging technique that combines high optical absorption contrast with diffraction-limited resolution of ultrasonic imaging, which allows deeper tissues to be viewed in living subjects. In photoacoustic imaging, short pulses of stimulating radiation are absorbed by tissues, resulting a subsequent thermal expansion and ultrasonic emission that can be detected by highly sensitive piezoelectric devices. However, many diseases, for example cancer, in their early stages, do not exhibit a natural photoacoustic contrast, therefore administering an external photoacoustic contrast agent is necessary. Owing to the strong light absorption characteristic [77], the CNTs can be utilized as photoacoustic contrast agents. De la Zerdaet. al, in the first time, applied SWNTs for in-vivo imaging of tumors in mice [103]. In this study, SWNTs was surface modified by PL-PEG and further conjugated with cyclic RGD peptides for targeting $\alpha v\beta 3$ integrin on cancer cells. Intravenous injection of these cyclic RGD functionalized CNTs to mice bearing tumors showed eight times stronger photoacoustic signal in the tumors than mice injected with non-targeted CNTs. This study suggested that photoacoustic imaging using targeted SWNTs could contribute to non-invasive in-vivo cancer imaging [103]. Similarly, in another study, SWNTs functionalized with antibody against $\alpha v\beta 3$ integrin for photoacoustic imaging of human glioblastoma tumors in nude mice [104].

5. Selected examples for preparation of carbon nanotube based therapeutics

In the above CNTs for drug delivery section, we described the functionalization of CNTs with drugs or targeting molecules. These preparations are usually applied directly via intravenous delivery route, which is the most widely used route of drug administration. Alternative to the intravenous drug administration, some other routes of drug administrations are also important for certain specific applications. Different formulations of CNT-based therapeutics have also been developed to suit the specific routes of administration. Here, we report several novel cases of CNTs applications for oral and transdermal delivery routes.

5.1. Carbon nanotube based therapeutics delivery using artificial cells: oral delivery

For many therapeutics, oral and targeted delivery are challenge. One way to deliver them at the targeted site is by novel methods of encapsulating them in polymeric artificial cells. Artificial cells are vesicles made by polymeric membranes. They can mimic certain functions of biological cells. The size of artificial cells ranged from nanometer to hundreds of micrometer [105]. The membranes of artificial cells are usually semi-permeable that allows for exchange of small molecules and prevention of passage of large substances across it. Up to date, artificial cells have been applied for encapsulation biologically active agents, including enzymes, hormones, drugs, even live bacteria cells for in-vivo delivery. Currently a couple of artificial cells have been applied clinically [106]. The advantages of artificial cells include protection of the cargos from immune elimination in the body, targeted delivery of cargos to desired sites and increasing cargo solubility [106]. In a first feasibility study, functionalized CNTs have been encapsulated in artificial cell made from biocompatible polymeric membrane for target specific delivery. The polymeric membrane was assembled with three layers of polymers, alginate-poly-L-lysine-alginate (APA), via electrostatic association (figure 6). Artificial cells protected CNT therapeutics from degradation by the harsh environments [107]. PH degradation profile of the polymeric membrane of artificial cells can be adjusted by composition of polymers, which allows the breakdown of the artificial cells and release of CNT therapeutics to desired sites. This system is ideal for oral delivery, and can be used for other delivery routes as well.

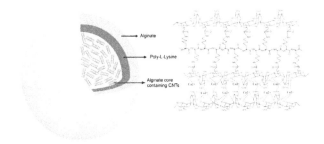

Figure 6. Alginate-poly-L-lysine-alginate (APA) microcapsules encapsulatingcarbon nanotubes. The calcium ions are responsible for cross-linking of the alginate monomeric units trapping the carbon nanotubes into the core of microcapsule [108].

5.2. Carbon nanotube based membrane: transdermal drug delivery

One of the most important areas of transdermal drug delivery (TDD) is in addiction treatment. Nicotine TDD has been widely used for smoking cessation programs. However, these traditional transdermal patches could not provide variable drug delivery rates. Some TDD has the capability to provide variable and programmable delivery rates, however, it needs a strong electric current across the human skin, which can cause serious skin irritation. Recently, the membranes prepared by functionalized CNTs have been employed in transdermal drug delivery. It has been shown that this CNT-membrane was very effective for treatment of drug abuse and addiction [109]. To prepare the CNT-membrane, MWNTs were firstly functionalized with negatively charged molecule containing sulphonate ($-SO_3^-$) to have a high charge density on the surface of CNTs, which is necessary to get efficient electro-osmosis pumping effect. By this functionalization strategy, CNT-membrane achieves dramatically fast flow through CNT cores, high charge density, and highly efficient electrophoretic pumping effect. These membranes were the integrated with a nicotine formulation to obtain switchable transdermal nicotine delivery rates on human skin (in vitro). The transdermal nicotine formulated CNT-membrane was able to successfully switch between high level and low level fluxes that coincide with therapeutic demand levels for nicotine cessation treatment. These programmable devices cause minimal skin irritation [109].

6. Potentials and limitations of carbon nanotubes in medicine

CNTs are being highly explored in the fields of targeted drug delivery, nanoscaffold for tissue engineering, biomedical imagining and detecting for disease treatment and health monitoring. The use of CNTs in drug delivery, detection and tissue engineering has shown the potentials to revolutionize medicine. CNTs affords for a large amounts of payloads for specific-targeting and drug delivery. With their intrinsic properties, the CNTs have potential for building-up multifunctional nanodevices for simultaneous therapeutic delivery and detection. Current cancer therapies (eg. radiation therapy, and chemotherapy) are usually painful and less efficient since they kill normal cells in addition to cancer cells, and therefore, producing adverse side effects and resistance. The CNT-based drug delivery systems have shown efficient tumor-targeting, and they can effectively kill cancer cells with a dosage lower than conventional drugs used, however significantly reduces side effects. Current CNT-based nanoscaffolds are very advantages for stem cells therapy in that they can modulate the stem cell growth and differentiation in a more controlled and desirable manner. Thus, CNTs have recently gained much interest in the field of medicine.

Although very useful, CNTs exist some limitations. Firstly, pristine CNTs, being metallic in their nature, are insoluble and they form large bundles or ropes in many solvents, including water and most solvents, so they cannot be used directly in biomedical applications. Much work has been done to prepare them for use in medicine. Secondly, the CNTs are not homogenous in their sizes (both diameters and lengths), which could be a problem for generation of reproducible results that allows evaluation of the biological activity relating to specific structures. Up to date, tremendous efforts have been put in surface functionalization of CNTs for use in medicine. This includes numerous effective methods for covalent or non-covalent modification of CNTs as to disperse them into aqueous solutions and to attach functional molecules for therapeutic applications. However, in terms of homegenecity of CNTs. Not much work has been done so far. We propose that attentions are needed to develop the methods for generation of CNTs with homogeneous size, which is very important for future clinical applications of CNT-based therapeutics.

As a novel nanomaterial of great potentials in medicine, the toxicology of CNTs has received much attention in recent years. Pristine CNTs are very light powders and they can enter the body through inhalation via the respiratory tract, ingestion via the gastrointestinal tract or, dermal absorption via the skin. Following entering, CNTs distribute rapidly in the central and peripheral nervous system, lymphatic and blood circulation and potentially cause toxic effects in a variety of tissues and organs that they reach, such as heart, spleen, kidney, bone marrow and liver, etc. Toxicity of CNTs has been evaluated in a variety of cell or animal models for assessing pulmonary, dermal and immune effects. However, the published results have not led to any consensus on the toxicity profile of pristine or functionalized SWNTs and MWNTs. Some investigators reported that pristine SWNTs that were purified by acid treatment demonstrated no acute toxicity, as opposed to non-purified CNTs, however, they induced reactive oxygen species (ROS) in human lung carcinoma epithelial A549 cells and NR8383 cells [110]. Others demonstrated that pristine SWNTs, either acid-treated or non-treated, were capable of increasing chromosome and DNA damage, and oxidative stress in macrophage cell lines [111, 112].

In contrast to raw or acid treated CNTs, the well-dispersed CNTs with high levels of surface functionalization can reduce the toxicity of MWNTs. One study demonstrated that taurine-MCWNTs in low and medium doses induced slight and recoverable pulmonary inflammation in mice, and are less toxic than raw MCWNTs [113]. This is supported by other studies indicating that the damage caused by non-PEGylated MWNTs is slightly more severe than that of PEGylated MWNTs [114]. Furthermore, administrations of high doses of PL-PEG functionalized SWNTs following intravenous injection did not lead to acute or chronic toxicity in nude mice, albeit SWNTs persisted within liver and spleen macrophages for 4 months in mice without apparent toxicity [115] and the SWNTs-PL-PEG were excreted from mice via the biliary and renal pathways [116]. It is hypothesized that the van der Waals forces on the surfaces of pristine CNTs cause hydrophobic interactions between CNTs, resulting in aggregation and network formation, which further induce prolonged toxicity. Thus, functionalization of CNTs overcomes the aggregate-forming surface properties of CNTs, and therefore, reduces toxicity.

7. Conclusions:

CNTs have exhibited diverse physical, chemical and mechanical properties suitable for a variety of applications. In last decade, biomedical applications of CNTs have undergone rapid progress. Their unique properties, such as, ultrahigh surface area, high aspect ratio, distinct optical properties have been applied to develop innovative, multi-functional CNT-based nanodevices for broad applications. This chapter have described the chemical and physical methods to prepare CNTs for used in medicine. With these methods, targeting molecules are attached on CNTs for targeted drug delivery, selective imaging, and other therapies. As a new type of nanomaterial, the toxicity of CNTs has been extensively investigated. To date, tremendous toxicity studies on CNTs have been published. However, the published data are inconsistent. The reason is that CNTs used in these studies vary in dispersion status, size and length of tubes, metal impurities and functionalization methods etc. Moreover, different analysis methods used in the evaluation CNTs toxicity studies also cause disparities. Despite these disparities, there is a broad agreement that well-dispersed CNTs have little or no toxicity both in-vitro and in-vivo, and therefore are useful for biomedical applications. Finally, an urgent need has been proposed for long-term studies on the absorption, deposition, metabolism and excretion (ADME) of CNTs. Only after the uncertainty on CNT toxicity is resolved, the CNT-based therapeutics can be possible applied clinically.

Acknowledgements

This work is partially supported by research grant to Satya Prakash from Canadian Institutes of Health Research (CIHR) (MOP 93641). W. Shao and L. Rodes acknowledges Doctor Training Award from Fonds de Research Sante (FRSQ). A. Paul acknowledges Post-Doctoral Award from FRSQ.

Author details

Wei Shao[1], Paul Arghya[1], Mai Yiyong[2], Laetitia Rodes[1] and Satya Prakash[1*]

*Address all correspondence to: satya.prakash@mcgill.ca

1 Biomedical Technology and Cell Therapy Research Laboratory, Department of Biomedical Engineering and Artificial cells and Organs Research Centre, Faculty of Medicine, McGill University, Canada

2 Department of Chemistry, McGill University, Canada

References

[1] Chen, Z., et al. (2011). Single-walled carbon nanotubes as optical materials for bio-sensing. *Nanoscale*, 3(5), 1949-1956.

[2] Bekyarova, E., et al. (2005). Electronic properties of single-walled carbon nanotube networks. *Journal of the American Chemical Society*, 5990-5995.

[3] Kostarelos, K., et al. (2007). Cellular uptake of functionalized carbon nanotubes is independent of functional group and cell type. *Nat Nanotechnol*, 2(2), 108-113.

[4] Ando, Y. (2010). Carbon nanotube: the inside story. *J Nanosci Nanotechnol*, 10(6), 3726-3738.

[5] Sang Won Kim, T.K, Yern Seung, Kim, Hong Soo, Choi, Hyeong, Jun Lim, Seung Jae, Yang, & Chong Rae, Park. (2012). Surface modifications for the effective dispersion of carbon nanotubes in solvents and polymers. *Carbon*, 50(1), 30.

[6] Tasis, D., et al. (2003). Soluble carbon nanotubes. *Chemistry*, 9(17), 4000-4008.

[7] O'Connell, M. J., et al. (2002). Band gap fluorescence from individual single-walled carbon nanotubes. *Science*, 297(5581), 593-596.

[8] Valerie, C., Moore, M. S. S., Erik, H., Haroz, Robert. H., & Hauge, Richard E. (2003). Smalley, Individually Suspended Single-Walled Carbon Nanotubes in Various Surfactants. *Nano Lett*, 3(10), 3.

[9] Strano, M. S., et al. (2003). The role of surfactant adsorption during ultrasonication in the dispersion of single-walled carbon nanotubes. *J Nanosci Nanotechnol*, 3(1-2), 81-86.

[10] Liu, Z., et al. (2009). Preparation of carbon nanotube bioconjugates for biomedical applications. *Nat Protoc*, 4(9), 1372-1382.

[11] Fukushima, T, et al. (2003). Molecular ordering of organic molten salts triggered by single-walled carbon nanotubes. *Science*, 300(5628), 2072-2074.

[12] Dong, B., et al. (2011). Dispersion of carbon nanotubes by carbazole-tailed amphiphilic imidazolium ionic liquids in aqueous solutions. *Journal of colloid and interface science*, 190-195.

[13] Fu, Q., & Liu, J. (2005). Effects of ionic surfactant adsorption on single-walled carbon nanotube thin film devices in aqueous solutions. *Langmuir : the ACS journal of surfaces and colloids*, 1162-1165.

[14] Tasis, D., et al. (2006). Chemistry of carbon nanotubes. *Chem Rev*, 106(3), 1105-1136.

[15] Petrov, P., et al. (2003). Noncovalent functionalization of multi-walled carbon nanotubes by pyrene containing polymers. *Chem Commun (Camb)* [23], 2904-2905.

[16] Li, H., et al. (2004). Selective interactions of porphyrins with semiconducting single-walled carbon nanotubes. *Journal of the American Chemical Society*, 1014-1015.

[17] Xianbao Wang, Y.L, Wenfeng, Qiu, & Daoben, Zhu. (2002). Immobilization of tetra-tert-butylphthalocyanines on carbon nanotubes: a first step towards the development of new nanomaterials. *J. Mater. Chem*, 12(6), 3.

[18] Kang, Y. K., et al. (2009). Helical wrapping of single-walled carbon nanotubes by water soluble poly(p-phenyleneethynylene). *Nano letters*, 9(4), 1414-1418.

[19] Zheng, M., et al. (2003). DNA-assisted dispersion and separation of carbon nanotubes. *Nature materials*, 338-342.

[20] Hu, H., et al. (2003). Nitric Acid Purification of Single-Walled Carbon Nanotubes. *The Journal of Physical Chemistry B*, 107(50), 13838-13842.

[21] Karousis, N., Tagmatarchis, N., & Tasis, D. (2010). Current progress on the chemical modification of carbon nanotubes. *Chem Rev*, 110(9), 5366-5397.

[22] Georgakilas, V., et al. (2002). Amino acid functionalisation of water soluble carbon nanotubes. *Chem Commun (Camb)* [24], 3050-3051.

[23] Ojima, I. (2008). Guided molecular missiles for tumor-targeting chemotherapy--case studies using the second-generation taxoids as warheads. *Accounts of chemical research*, 108-119.

[24] Krauss, W. C., et al. (2000). Emerging antibody-based HER2 (ErbB-2/neu) therapeutics. *Breast Dis*, 11, 113-124.

[25] Abramson, J. S., & Shipp, M. A. (2005). Advances in the biology and therapy of diffuse large B-cell lymphoma: moving toward a molecularly targeted approach. *Blood*, 1164-1174.

[26] Garanger, E., Boturyn, D., & Dumy, P. (2007). Tumor targeting with RGD peptide ligands-design of new molecular conjugates for imaging and therapy of cancers. *Anticancer Agents Med Chem*, 7(5), 552-558.

[27] Corti, A., & Curnis, F. (2011). Tumor Vasculature Targeting Through NGR Peptide-Based Drug Delivery Systems. *Current Pharmaceutical Biotechnology*, 1128-1134.

[28] Lu, Y., & Low, P. S. (2002). Folate-mediated delivery of macromolecular anticancer therapeutic agents. *Advanced drug delivery reviews*, 675-693.

[29] Hermanson, G. T. (2008). Bioconjugate Techniques,2nd Edition. Academic Press, Inc., 1202, pages.

[30] Liu, Z., et al. (2007). Supramolecular chemistry on water-soluble carbon nanotubes for drug loading and delivery. *ACS nano*, 50-56.

[31] Ji, Z., et al. (2012). Targeted therapy of SMMC-7721 liver cancer in vitro and in vivo with carbon nanotubes based drug delivery system. *Journal of colloid and interface science*, 143-149.

[32] Wang, L., et al. (2011). Synergistic enhancement of cancer therapy using a combination of docetaxel and photothermal ablation induced by single-walled carbon nanotubes. *Int J Nanomedicine*, 6, 2641-2652.

[33] Taghdisi, S. M., et al. (2011). Reversible targeting and controlled release delivery of daunorubicin to cancer cells by aptamer-wrapped carbon nanotubes. *European journal of pharmaceutics and biopharmaceutics official journal of Arbeitsgemeinschaft fur Pharmazeutische Verfahrenstechnik e.*, 77, 200-206.

[34] Lu, Y. J., et al. (2012). Dual targeted delivery of doxorubicin to cancer cells using folate-conjugated magnetic multi-walled carbon nanotubes. *Colloids Surf B Biointerfaces*, 89, 1-9.

[35] Lay, C. L., et al. (2010). Delivery of paclitaxel by physically loading onto poly(ethylene glycol) (PEG)-graft-carbon nanotubes for potent cancer therapeutics. *Nanotechnology*, 21(6), 065101.

[36] Bhirde, A. A., et al. (2009). Targeted killing of cancer cells in vivo and in vitro with EGF-directed carbon nanotube-based drug delivery. *ACS nano*, 307-316.

[37] Liu, Z., et al. (2008). Drug delivery with carbon nanotubes for in vivo cancer treatment. *Cancer research*, 68(16), 6652-6660.

[38] Tsavaris, N. B., & Kosmas, C. (1998). Risk of severe acute hypersensitivity reactions after rapid paclitaxel infusion of less than 1-h duration. *Cancer chemotherapy and pharmacology*, 509-511.

[39] Sendo, T., et al. (2005). Incidence and risk factors for paclitaxel hypersensitivity during ovarian cancer chemotherapy. *Cancer chemotherapy and pharmacology*, 91-96.

[40] Zhang, X., et al. (2009). Targeted delivery and controlled release of doxorubicin to cancer cells using modified single wall carbon nanotubes. *Biomaterials*, 30(30), 6041-6047.

[41] Yang, F, et al. (2011). Magnetic functionalised carbon nanotubes as drug vehicles for cancer lymph node metastasis treatment. *Eur J Cancer*, 47(12), 1873-1882.

[42] Dhar, S., et al. (2008). Targeted single-wall carbon nanotube-mediated Pt(IV) prodrug delivery using folate as a homing device. *Journal of the American Chemical Society*, 130(34), 11467-11476.

[43] Sun, X., & Zhang, N. (2010). Cationic polymer optimization for efficient gene delivery. *Mini reviews in medicinal chemistry*, 108-125.

[44] Shao, W., et al. (2012). A novel polyethyleneimine-coated adeno-associated virus-like particle formulation for efficient siRNA delivery in breast cancer therapy: preparation and in vitro analysis. *Int J Nanomedicine*, 7, 1575-1586.

[45] Yang, F., et al. (2009). Pilot study of targeting magnetic carbon nanotubes to lymph nodes. *Nanomedicine*, 317-330.

[46] Nunes, A, et al. (2010). Hybrid polymer-grafted multiwalled carbon nanotubes for in vitro gene delivery. *Small*, 6(20), 2281-2291.

[47] Pantarotto, D., et al. (2004). Functionalized carbon nanotubes for plasmid DNA gene delivery. *Angew Chem Int Ed Engl*, 43(39), 5242-5246.

[48] Singh, R., et al. (2005). Binding and condensation of plasmid DNA onto functionalized carbon nanotubes: toward the construction of nanotube-based gene delivery vectors. *Journal of the American Chemical Society*, 4388-4396.

[49] Wang, X., et al. (2009). Knocking-down cyclin A(2) by siRNA suppresses apoptosis and switches differentiation pathways in K562 cells upon administration with doxorubicin. *PloS one*, 4(8), e6665.

[50] Zhang, Z., et al. (2006). Delivery of telomerase reverse transcriptase small interfering RNA in complex with positively charged single-walled carbon nanotubes suppresses tumor growth. *Clinical cancer research : an official journal of the American Association for Cancer Research*, 4933-4939.

[51] Wang, X., Ren, J., & Qu, X. (2008). Targeted RNA interference of cyclin A2 mediated by functionalized single-walled carbon nanotubes induces proliferation arrest and apoptosis in chronic myelogenous leukemia K562 cells. *ChemMedChem*, 3(6), 940-945.

[52] Yang, R., et al. (2006). Single-walled carbon nanotubes-mediated in vivo and in vitro delivery of siRNA into antigen-presenting cells. *Gene therapy*, 13(24), 1714-1723.

[53] Liu, Y., et al. (2005). Polyethylenimine-grafted multiwalled carbon nanotubes for secure noncovalent immobilization and efficient delivery of DNA. *Angew Chem Int Ed Engl*, 44(30), 4782-4785.

[54] Liu, Z, et al. (2007). siRNA delivery into human T cells and primary cells with carbon-nanotube transporters. *Angew Chem Int Ed Engl*, 46(12), 2023-2027.

[55] Kam, N. W., Liu, Z., & Dai, H. (2005). Functionalization of carbon nanotubes via cleavable disulfide bonds for efficient intracellular delivery of siRNA and potent gene silencing. *Journal of the American Chemical Society*, 127(36), 12492-12493.

[56] Bartholomeusz, G., et al. (2009). In Vivo Therapeutic Silencing of Hypoxia-Inducible Factor 1 Alpha (HIF-1alpha) Using Single-Walled Carbon Nanotubes Noncovalently Coated with siRNA. *Nano Res*, 2(4), 279-291.

[57] Paul, A., et al. (2009). Microencapsulated stem cells for tissue repairing: implications in cell-based myocardial therapy. *Regen Med*, 4(5), 733-745.

[58] Harrison, B. S., & Atala, A. (2007). Carbon nanotube applications for tissue engineering. *Biomaterials*, 28(2), 344-353.

[59] Chao, T. I., et al. (2010). Poly(methacrylic acid)-grafted carbon nanotube scaffolds enhance differentiation of hESCs into neuronal cells. *Adv Mater*, 22(32), 3542-3547.

[60] Mooney, E, et al. (2008). Carbon nanotubes and mesenchymal stem cells: biocompatibility, proliferation and differentiation. *Nano Lett*, 8(8), 2137-2143.

[61] Martinelli, V, et al. (2012). Carbon nanotubes promote growth and spontaneous electrical activity in cultured cardiac myocytes. *Nano Lett*, 12(4), 1831-1838.

[62] Stout, D. A., Basu, B., & Webster, T. J. (2011). Poly(lactic-co-glycolic acid): carbon nanofiber composites for myocardial tissue engineering applications. *Acta Biomater*, 7(8), 3101-3112.

[63] Oh, S, et al. 2009. Stem cell fate dictated solely by altered nanotube dimension. *Proc Natl Acad Sci U S A*, 106(7), 2130-2135.

[64] Tran, P. A., Zhang, L., & Webster, T. J. (2009). Carbon nanofibers and carbon nanotubes in regenerative medicine. *Adv Drug Deliv Rev*, 61(12), 1097-1114.

[65] Park, J., et al. (2012). Synergistic control of mesenchymal stem cell differentiation by nanoscale surface geometry and immobilized growth factors on TiO2 nanotubes. *Small*, 8(1), 98-107.

[66] Lai, M., et al. (2011). Surface functionalization of TiO2 nanotubes with bone morphogenetic protein 2 and its synergistic effect on the differentiation of mesenchymal stem cells. *Biomacromolecules*, 12(4), 1097-1105.

[67] Hu, Y., et al. (2012). TiO2 nanotubes as drug nanoreservoirs for the regulation of mobility and differentiation of mesenchymal stem cells. *Acta Biomater*, 8(1), 439-448.

[68] Connick, P., Patani, R., & Chandran, S. (2011). Stem cells as a resource for regenerative neurology. *Pract Neurol*, 11(1), 29-36.

[69] Levenberg, S., et al. (2005). Neurotrophin-induced differentiation of human embryonic stem cells on three-dimensional polymeric scaffolds. *Tissue Eng*, 11(3-4), 506-512.

[70] Jan, E., & Kotov, N. A. (2007). Successful differentiation of mouse neural stem cells on layer-by-layer assembled single-walled carbon nanotube composite. *Nano Lett*, 7(5), 1123-1128.

[71] Mazzatenta, A., et al. (2007). Interfacing neurons with carbon nanotubes: electrical signal transfer and synaptic stimulation in cultured brain circuits. *J Neurosci*, 27(26), 6931-6936.

[72] Nho, Y., et al. (2010). Adsorption of mesenchymal stem cells and cortical neural stem cells on carbon nanotube/polycarbonate urethane. *Nanomedicine (Lond*, 5(3), 409-417.

[73] Zhang, S. C., et al. (2001). In vitro differentiation of transplantable neural precursors from human embryonic stem cells. *Nat Biotechnol*, 19(12), 1129-1133.

[74] Chao, T. I., et al. (2009). Carbon nanotubes promote neuron differentiation from human embryonic stem cells. *Biochem Biophys Res Commun*, 384(4), 426-430.

[75] Kim, J. A., et al. (2012). Regulation of morphogenesis and neural differentiation of human mesenchymal stem cells using carbon nanotube sheets. *Integr Biol (Camb*, 4(6), 587-594.

[76] Chen, C. S., et al. (2012). Human stem cell neuronal differentiation on silk-carbon nanotube composite. *Nanoscale Res Lett*, 7(1), 126.

[77] Berber, S., Kwon, Y. K., & Tomanek, D. (2000). Unusually high thermal conductivity of carbon nanotubes. *Phys Rev Lett*, 84(20), 4613-4616.

[78] Marches, R., et al. (2009). Specific thermal ablation of tumor cells using single-walled carbon nanotubes targeted by covalently-coupled monoclonal antibodies. *Int J Cancer*, 125(12), 2970-2977.

[79] Gannon, C. J., et al. (2007). Carbon nanotube-enhanced thermal destruction of cancer cells in a noninvasive radiofrequency field. *Cancer*, 110(12), 2654-2665.

[80] Wang, C. H., et al. (2011). Photothermolysis of glioblastoma stem-like cells targeted by carbon nanotubes conjugated with CD133 monoclonal antibody. *Nanomedicine*, 7(1), 69-79.

[81] Chakravarty, P., et al. (2008). Thermal ablation of tumor cells with antibody-functionalized single-walled carbon nanotubes. *Proceedings of the National Academy of Sciences of the United States of America*, 105(25), 8697-8702.

[82] Xiao, Y., et al. (2009). Anti-HER2 IgY antibody-functionalized single-walled carbon nanotubes for detection and selective destruction of breast cancer cells. *BMC cancer*, 9, 351.

[83] Kam, N. W., et al. (2005). Carbon nanotubes as multifunctional biological transporters and near-infrared agents for selective cancer cell destruction. *Proceedings of the National Academy of Sciences of the United States of America*, 102(33), 11600-11605.

[84] Burke, A. R., et al. (2012). The resistance of breast cancer stem cells to conventional hyperthermia and their sensitivity to nanoparticle-mediated photothermal therapy. *Biomaterials*, 33(10), 2961-2970.

[85] Liu, Z., et al. (2007). In vivo biodistribution and highly efficient tumour targeting of carbon nanotubes in mice. *Nat Nanotechnol*, 2(1), 47-52.

[86] Burke, A., et al. (2009). Long-term survival following a single treatment of kidney tumors with multiwalled carbon nanotubes and near-infrared radiation. *Proc Natl Acad Sci U S A*, 106(31), 12897-12902.

[87] Ghosh, S., et al. (2009). Increased heating efficiency and selective thermal ablation of malignant tissue with DNA-encased multiwalled carbon nanotubes. *ACS Nano*, 3(9), 2667-2673.

[88] Singh, S. K., et al. (2003). Identification of a cancer stem cell in human brain tumors. *Cancer Res*, 63(18), 5821-5828.

[89] Lanzetta, G., et al. (2003). Temozolomide in radio-chemotherapy combined treatment for newly-diagnosed glioblastoma multiforme: phase II clinical trial. *Anticancer Res*, 23(6D), 5159-5164.

[90] Zeppernick, F., et al. (2008). Stem cell marker CD133 affects clinical outcome in glioma patients. *Clin Cancer Res*, 14(1), 123-129.

[91] Mocan, T, & Iancu, C. (2011). Effective colon cancer prophylaxis in mice using embryonic stem cells and carbon nanotubes. *Int J Nanomedicine*, 6, 1945-1954.

[92] Xie, S. S., et al. (2000). Mechanical and physical properties on carbon nanotube. *Journal of Physics and Chemistry of Solids*, 61(7), 1153-1158.

[93] Tasis, D., et al. (2008). Diameter-selective solubilization of carbon nanotubes by lipid micelles. *J Nanosci Nanotechnol*, 8(1), 420-423.

[94]   Strano, M. S., & Jin, H. (2008). Where is it heading? Single-particle tracking of single-walled carbon nanotubes. *ACS nano*, 2(9), 1749-1752.

[95]   Jin, H., Heller, D. A., & Strano, M. S. (2008). Single-particle tracking of endocytosis and exocytosis of single-walled carbon nanotubes in NIH-3T3 cells. *Nano letters*, 8(6), 1577-1585.

[96]   Keren, S., et al. (2008). Noninvasive molecular imaging of small living subjects using Raman spectroscopy. *Proceedings of the National Academy of Sciences of the United States of America*, 105(15), 5844-5849.

[97]   Zavaleta, C., et al. (2008). Noninvasive Raman spectroscopy in living mice for evaluation of tumor targeting with carbon nanotubes. *Nano letters*, 8(9), 2800-2805.

[98]   Liu, Z., et al. (2010). Multiplexed Five-Color Molecular Imaging of Cancer Cells and Tumor Tissues with Carbon Nanotube Raman Tags in the Near-Infrared. *Nano Res*, 3(3), 222-233.

[99]   Welsher, K., et al. (2008). Selective probing and imaging of cells with single walled carbon nanotubes as near-infrared fluorescent molecules. *Nano letters*, 8(2), 586-590.

[100] Welsher, K., et al. (2009). A route to brightly fluorescent carbon nanotubes for near-infrared imaging in mice. *Nat Nanotechnol*, 4(11), 773-780.

[101] Hanlon, E. B., et al. (2000). Prospects for in vivo Raman spectroscopy. *Physics in medicine and biology*, 45(2), R1-59.

[102] Karmakar, A., et al. (2011). Raman spectroscopy as a detection and analysis tool for in vitro specific targeting of pancreatic cancer cells by EGF-conjugated, single-walled carbon nanotubes. *J Appl Toxicol*.

[103] De la Zerda, A., et al. (2008). Carbon nanotubes as photoacoustic molecular imaging agents in living mice. *Nat Nanotechnol*, 3(9), 557-562.

[104] Xiang, L., et al. (2009). Photoacoustic molecular imaging with antibody-functionalized single-walled carbon nanotubes for early diagnosis of tumor. *J Biomed Opt*, 14(2), 021008.

[105] Chang, T. M. (2004). Artificial cell bioencapsulation in macro, micro, nano, and molecular dimensions: keynote lecture. *Artificial cells, blood substitutes, and immobilization biotechnology*, 32(1), 1-23.

[106] Prakash, S. (2007). Artificial cells, cell engineering and therapy:. *Woodhead*.

[107] Kulamarva, A., et al. (2009). Microcapsule carbon nanotube devices for therapeutic applications. *Nanotechnology*, 20(2), 025612.

[108] Prakash, S., et al. (2011). Polymeric nanohybrids and functionalized carbon nanotubes as drug delivery carriers for cancer therapy. *Advanced drug delivery reviews*, 63(14-15), 1340-1351.

[109] Wu, J., et al. (2010). Programmable transdermal drug delivery of nicotine using carbon nanotube membranes. *Proceedings of the National Academy of Sciences of the United States of America*, 107(26), 11698-11702.

[110] Pulskamp, K., Fau-Diabate, S., Diabate, H. F., Fau-Krug, S., & Krug, H. F. Carbon nanotubes show no sign of acute toxicity but induce intracellular reactive oxygen species in dependence on contaminants. 0378-4274, Print.

[111] Migliore, L., Fau-Saracino, D., et al. Carbon nanotubes induce oxidative DNA damage in RAW 264.7 cells. 1098-2280, Electronic.

[112] Kagan Ve Fau- Tyurina, Y.Y, et al. Direct and indirect effects of single walled carbon nanotubes on RAW 264.7 macrophages: role of iron. 0378-4274, Print.

[113] Wang, X., Fau-Zang, J. J., et al. Pulmonary toxicity in mice exposed to low and medium doses of water-soluble multi-walled carbon nanotubes. 1533-4880, Print.

[114] Zhang, D., Fau-Deng, X., et al. Long-term hepatotoxicity of polyethylene-glycol functionalized multi-walled carbon nanotubes in mice. 1361-6528, Electronic.

[115] Schipper, Ml., Fau-Nakayama-Ratchford, N., et al. A pilot toxicology study of single-walled carbon nanotubes in a small sample of mice. 1748-3395, Electronic.

[116] Liu, Z., Fau-Davis, C., et al. Circulation and long-term fate of functionalized, biocompatible single-walled carbon nanotubes in mice probed by Raman spectroscopy. 1091-6490, Electronic.

Chapter 3

Latest Advances in Modified/Functionalized Carbon Nanotube-Based Gas Sensors

Enid Contés-de Jesús, Jing Li and Carlos R. Cabrera

Additional information is available at the end of the chapter

1. Introduction

A gas sensor is a device that when exposed to gaseous species, is able to alter one or more of its physical properties, so that can be measured and quantified, directly or indirectly. These devices are used for applications in homeland security, medical diagnosis, environmental pollution, food processing, industrial emission, public security, agriculture, aerospace and aeronautics, among others. Desirable characteristics of a gas sensor are selectivity for different gases, sensitivity at low concentrations, fast response, room temperature operation (some applications may require high temperature), low power consumption, low-cost, low maintenance and portability. Traditional techniques like gas chromatography (GC), GC coupled to mass spectrometry (GC-MS), Fourier transmission infrared spectroscopy (FTIR) and atomic emission detection (AED) provide high sensitivity, reliability and precision, but they are also bulky, time consuming, power consuming, operate at high temperature, and the high maintenance and requirement of trained technicians translate in high costs. In an effort to overcome those disadvantages, research in the area has been focused on the search for functional sensing materials.

Carbon nanotubes (CNTs) have been have been focus of intense research as alternative sensing material because of their attractive characteristics like chemical, thermal and mechanical stability, high surface area, metallic and semi-conductive properties and functionalization capability [1]. CNTs are graphene sheets rolled in a tubular fashion. Different types of CNTs can be synthesized: single walled carbon nanotubes (SWCNT), double wall carbon nanotubes (DWCNTs) and multi walled carbon nanotubes (MWCNT).

The publication of the first CNT-based sensor for NH_3 and NO_2 detection using an individual semiconducting SWCNT [2] triggered the research activity in this area. Pristine CNTs have shown to be chemically inactive to gas molecules in general. However, their modification/functionalization capability has been exploited throughout the last years, especially for the development of devices with enhanced selectivity and sensitivity for the room temperature detection of a wide variety of gases. Numerous articles and reviews focused on different aspect of CNTs-based sensors and summarizing their progress and potential have been published throughout the years. Some of them are listed in references [3-34]. A most recent review [35] addressing the technological and commercial aspects of CNTs sensors presents evidence of the continuous active research in the area and that they have real potential to complement or substitute current technologies.

This chapter presents a summary of selected original research articles that have been published between 2010 and present in which the main subject are modified/functionalized SWCNTs, DWCNTs and MWCNTs and their use as gas sensing material. The majority of the references included in this chapter content are based on experimental results. However, theoretical studies based on computational science are also included because of their importance in the study of CNT-based sensors. The use of different methods of calculations and simulation has been useful to design new structures and materials and to study, evaluate and predict the interactions and adsorption energies between those materials and gaseous molecules. First, we present current research activities on pristine CNTs-based sensors and the different approaches used to improve their sensitivity and selectivity without modifying the CNTs structure, followed by the review of CNTs modified with conducting polymers, metallic nanoparticles (NPs), nanostructured oxides and sidewall modification, doping and others. Different modification/functionalization techniques like chemical deposition, plasma, sputtering and electrodeposition are discussed. Gas sensors based on changes of electrical conductivity caused by adsorption of gas molecules (resistors) are the most common sensor type discussed in this review. Other sensing platforms like surface acoustic wave (SAW) and quartz microbalance (QMB)are also included.

2. Unmodified carbon nanotubes

Pristine CNTs are known for their high stability because of their strong sp^2 carbon-carbon bonds and thus insensitive when used as sensing material for certain gases. However, the detection of NO, NO_2 and NH_3 has been previously reported. In order to improve their sensitivity and recovery time, different approaches like dispersion techniques to debundling the CNTs ropes, humidity assisted detection, application of an electric field, continuous use of ultraviolet (UV) light and even separation of semi-conductive types from conductive have been reported. The detection of NO, NO_2 and NH_3, as well as other gases like formaldehyde and dimethyl methylphosphonate (DMMP) with pristine SWCNTs are discussed in this section.

A MWCNTs based sensor was used for the detection of 50 ppm of nitrogen monoxide (NO) [36]. With the purpose to increase the sensitivity, an electric field was applied between two copper plates as electrodes, one of them containing MWCNTs-silicon wafer. It was found that when a positive potential was applied to the copper plate and the negative potential applied to the copper plate containing the MWCNTs-sensor, NO, being an electron acceptor, moves to the electron enriched zone, which is the one containing the MWCNTs-sensor and thus enhancement in the sensitivity is observed. The stronger the applied electric field, the better the sensitivity. However, applying a negative electric field was applied to the copper plate and a positive potential was applied to the copper plate containing the MWCNTs-sensor, the NO molecules moved away from the MWCNTs sensor and thus a decrease in the sensitivity is observed. The more negative the applied electric field, the lower the sensitivity of the sensor. Recovery of the sensors was achieved by applying reverse potential from the one used to perform the gas sensing experiments.

Cava, et al. proposed the use of a homogeneous film of MWCNTs prepared by the self-assembly technique and use it as an active layer for an oxygen gas sensor with increased sensitivity [37]. When the sensors were exposed to 10% O_2 in Nitrogen at 160 °C, the electrical resistance decreased and showed a better oxygen sensitivity when compared to sensors prepared under the same condition but using the drop-cast method. The reason for this is that the self-assembly technique provides a better distribution of the nanotubes and thus promoting a better gas adsorption between nanotubes (inter-tube contact).

The high van der walls attraction between CNTs causes them to remain in bundles or agglomerated. This can represent a problem for their application as gas sensors because it results in less adsorption/interaction (binding) sites, which translates in less sensitivity. Considering this, different dispersion techniques were used by Ndiaye and coworkers for the preparation of CNTs based sensors for NO_2 detection [38]. SWCNTs were dispersed in a surfactant, sodium dodecylbenzene sulfonate (NaDDBs) and an organic solvent, chloroform ($CHCl_3$), drop-casted in IDEs and tested for the detection of 50, 100, 120, 200 ppb of NO_2 at 80 °C. Sensors prepared with SWCNTs dispersed in NaDDBs showed better sensitivity than those with SWCNTs dispersed in chloroform. The explanation to this is that the surfactant was more effective in debundle the SWCNTs than the organic solvent. It was stated that even though the surfactant was not completely removed after several rinsing steps and heating treatment at 150°C, it does not have significant effect on the electronic behavior of the sensor. Both surfactant-dispersed and organic solvent-dispersed samples showed a decrease in resistance with increasing temperature, which demonstrate the semi-conductive behavior of the SWCNTs and thus no effect of the solvent.

A SWCNT-based gas sensor selective for NO_2 and SO_2 at room temperature and ambient pressure was developed by Yao et al. [39]. High sensitivity and selectivity for 2 ppm of each gas was achieved by controlling the humidity levels. For instance, at low humidity levels, the sensors showed to be selective for NO_2 and insensitive to SO_2. At high humidity levels (92%), both gases were detected. However, NO_2 showed a decrease in resistance and SO_2 showed an increase in resistance.

Continuous in situ UV illumination on SWCNTs during gas sensing experiments was used to enhance the sensor's overall performance in the detection of NO, NO_2 and NH_3 (Figure 1) [40]. Changes in conductance ($\Delta G/G_0$) as function of time were recorded and used to prepare calibration curves in order to determine sensors sensitivity. It was found that the continuous exposure of the sensors to UV light under inert atmosphere (N_2, Ar) regenerating the surface, therefore, enhances their sensitivity for the detection of NO and NO_2. Linear responses were achieved at low concentrations and up to 50 ppm. Detection limits (DL), derived from the noise of the baseline and the slope obtained from the calibration curve, were found to be as low as 590 parts per quadrillion (ppq) and 1.51ppt for NO and NO_2, respectively. For NH_3 it was found not only that the *in situ* UV illumination reverses the direction of the changes in conductance, but it was also confirmed that it helps to improve the DL from 5.67 ppm to 27.8 ppt when tested under identical conditions. The achieved DL outperformed by several orders of magnitude the sensitivity of other CNTs-based NO, NO_2 and NH_3 sensors that have been previously reported. This is attributed to the UV light inducing surface regeneration and actively removal of all gases adsorbed on SWCNTs surface. It is worth noticing that this *in situ* cleaning with continuous UV-light exposure without device degradation was just achieved under inert atmospheres.

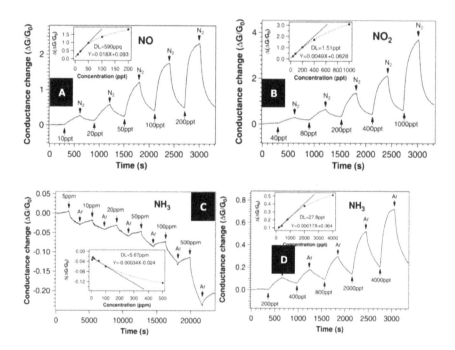

Figure 1. SWCNTs-sensor responses to (A) NO (10 – 200ppt) under *in situ* UV illumination, (B) NO_2 (40 – 1000 ppt) under *in situ* UV illumination, (C) NH_3 (5 – 500 ppm) without *in situ* UV illumination, and (D) NH_3 (200 – 4000 ppt) under *in situ* UV illumination. (From Chen *et al.* [40]. Copyright © 2012, with permission from Nature Publishing Group.)

Battie and coworkers used sorted semi-conducting SWCNTs as sensing film for the detection of NO_2 and NH_3 [41]. The density gradient ultracentrifugation (DGU) technique was used to separate semi-conducting from as produced SWCNTs. Films of as produced and sorted semi-conducting SWCNTs were exposed to NO_2 and NH_3 in air. Both films showed decrease in resistance and increase in resistance as exposed to NO_2 and NH_3, respectively. Full recovery was achieved by applying heat after NH_3 exposure and vacuum after NO_2 exposure. However, semi-conducting SWCNTs films were more sensitive for NH_3 than to NO_2 at different concentrations at ppm level.

Horrillo *et al.* used SWCNTs films for the room temperature detection of Chemical Warfare Agents (CWA) [42]. Changes in resistance were measured as samples were exposed to simulants of CWA at different ppm levels, DL of 0.01 ppm, 0.1 ppm and 50 ppm were achieved for DMMP, dipropylene glycol methyl ether (DPGME) and dimethylacetamide (DMA), respectively. The most remarkable advantage is that the sensors perform better and at more sensitive at room temperature, when tested at different temperatures.

3. Surface modified carbon nanotubes

CNTS modified with different functional groups have been used for the development of sensors for detection of volatile organic compounds (VOCs) in the environment as well as in exhaled breath. For the detection of VOCs in air, Wang *et al.* worked in the preparation of a sensor array based on MWCNTs covalently modified with different functional groups like propargyl, allyl, alkyltriazole, thiochain, thioacid, hexafluoroisopropanol (HFIP) [43] and Shrisat, *et al.* reported another one based on SWCNTs modified with different porphyrins (organic macrocyclic compounds) likeoctaethyl porphyrin (OEP), ruthenium OEP (RuOEP), iron OEP (FeOEP), tetraphenylporphyrin (TPP), among others [44]. Penza *et al.* also worked in the modification of MWCNTs with TPP for the room temperature detection of VOCs [45]. In this case, the TTP contained two different metals, Zn (CNT:ZnTPP) and Mn (CNT:MnTPP). Sensors were exposed to ethanol, acetone, ethylacetate, toluene and Triethylamine at ppm levels and all showed increase in resistance when exposed to the different gases. CNT: MnTPP showed the highest sensitivity towards all gases with respect to unmodified CNTs but for triethylamine and CNT: ZnTPP was more sensitive to ethylacetate.

Two different CNT-based sensor arrays have been reported for the detection and pattern recognition of VOCs present in exhaled breath samples for medical diagnosis, Tisch *et al.* presented a sensor array containing different nanomaterials including organically functionalized random networks of SWCNTs for the detection of VOCs related to Parkinson disease and that were present in exhaled breath collected from rats [46]. Ionescu and coworkers reported a sensor array based on bilayers of SWCNTS and polycyclic aromatic hydrocarbons (PAH) for the detection of multiple sclerosis in exhaled human breath [47]. In general, the incorporation of organic functional groups provided not only enhanced sensitivity but also provided better selectivity for each gas when compared to pristine CNTs. The use of statistical techniques like principal component analysis (PCA), discriminant factor analysis (DFA) and linear discriminant analysis (LDA) was possible to determine the discrimination capability of the sensors toward each VOC.

SWCNTs were functionalized with tetrafluorohydroquinone (TFQ) at the room temperature for detection of dimethyl methylphosponate (DMMP) at parts per trillion (ppt) levels (Fig-

ure 2) [48]. The conductance of the TFQ-SWCNTs samples increased as function of concentration when exposed to DMMP in N_2 in a concentration range from 20ppt to 5.4ppb.
Sensors showed fast response and ultra sensitivity down to 20ppt when compared to unmodified SWCNTs sensor, which had a DL of nearly 1 ppm. The presence of TFQ clearly
showed to improve the sensitivity and this is because it provides additional binding sites
thru hydrogen bonds between hydroxyl groups in TQF and DMMP. In addition, TQF tailors
the electronic properties of SWCNTs via hole doping.

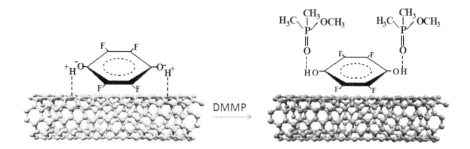

Figure 2. Representation of the possible sensing mechanism of DMMP with TFQ functionalized SWCNTs. (From Wei *et al.* [48]. Copyright © 2011, with permission from Institute of Physics Publishing.)

Wang *et al.* deposited a uniform network of SWCNT with carboxylic groups (-COOH) on a
flexible poly (dimethyldiallylammonium chloride) (PDDA) modified polyimide (PI) substrate for DMMP detection at room temperature [49]. Changes in resistance as function of
time were measured as the sensors were exposed to DMMP in N_2 at a concentration range of
1-40 ppm. Calibration curve showed a [17] linear increase of resistance as function of concentration and responses were found to be fast, stable and reproducible. Sensors showed to
be selective to DMMP in presence of other volatile organic vapors like methanol, xylene and
hexane, among others. Changing the carrier gas from N_2 to air caused the responses to
DMMP to be lower which might be due to influence of oxygen and humidity contained in
air. The apparent enhanced performance of this SWCNTs (-COOH) flexible sensor when
compared to sensors prepared on Si/SiO$_2$ rigid substrates is attributed to the presence of
PDDA, which is a polymer that absorbs DMMP. There is no information on the adsorption
of DMMP by PDDA and its possible effect in the recovery of the sensors.

MWCNTs were oxidized with KMnO4 to add oxygenated functional groups, mainly (-
COOH) for the detection of organic vapors [50]. Oxidized MWCNTs in form of a bucky paper were exposed to different concentrations of acetone. Variations in electrical resistance
were recorded for both unmodified and oxidized MWCNTs. Oxidized MWCNTs showed
higher sensitivity to acetone (2.3 vol. %) than unmodified ones, and good selectivity when
sensing other oxygen containing vapors such as diethyl ether and methanol. The sensors also showed complete reversibility and high reproducibility for all tested vapors.

MWCNTs were chemically treated with acid to obtain hydroxyl groups (OH) and used as sensing material for humidity sensors [51]. Changes in resistance were measured as the RH was varied from 11% to 98%. It was observed that the resistance increased as sensors were exposed to the different humidity levels. It was found that acid treated SWCNTS were more sensitive to humidity than pristine MWCNTs. The higher sensitivity of Acid treated SWCNTs is attributed to their higher surface and thus more adsorption sites that result from the acid treatment. Sensors showed fast response and to be stable. As with most humidity sensors, the recovery time was longer than response time due to slow desorption of water molecules.

Purified MWCNTs were treated with oxygen plasma or fluorine plasma and used for the detection of ethanol (Figure 3) [52]. Changes in resistance as function of time were recorded for the sensors when exposed of 50-500 ppm of ethanol vapor in air. Samples treated with oxygen plasma for 30 sec and with fluorine plasma for 60 sec showed the highest sensitivity to 100 ppm of ethanol, compared to pristine MWCNTS and other oxygen and fluorine plasma treated for different duration time. However, fluorine plasma treated samples showed the better sensitivity and reduced response and recovery time. The improvement of the fluorine plasma treated samples is explained by the difference in electronegativity between oxygen and fluorine.

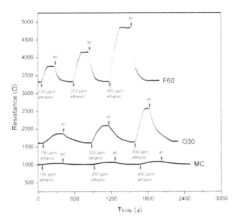

Figure 3. Responses of pristine (MC), oxygen plasma treated (O30) and fluorine plasma treated (F60) MWCNTs to 100-400 ppm of ethanol. (From Liu et al. [52]Copyright © 2012, with permission from Elsevier.)

Thermally fluorinated MWCNTs (TFC) were used for NO gas detection at room temperature [53]. The effect of thermal fluorination process was performed at various temperatures (100 -1000 °C) and 200 °C was found to be the optimum fluorination temperature. TFC samples prepared at temperatures higher than 200 °C showed a decrease of the fluorine functional groups, and even fluorine-assisted pyrolysis and fluorine-induced reorientation of the MWCNTs structure occurred at 1000 °C. TFC prepared at 200 °C showed high sensitivity, stability, reproducibility and full recovery, when their gas sensing properties were evaluated towards the detection of 50 ppm NO in dry air. Interestingly, the presence of fluorine reverses the electron transfer process, when compared to pristine MWCNTs, allowing them to go from NO to MWCNTs and thus causing an increase in resistance. The fluorination not only helped to enhance the sensitivity but also made the sensors insensitive to humidity changes.

MWCNTs were modified with amino groups for the detection of formaldehyde at ppb level [54]. Changes in resistance as function of time were measured as the sensors were exposed to formaldehyde in a concentration range between 20 and 200 ppb. Sensors containing MWCNTs with higher amino group content (18%) were 2.4 and 13 times more sensitive to formaldehyde than samples containing 5% amino groups and pristine CNTs, respectively. Short response times are due to a chemical reaction between the aldehyde and amino group. For the same reason, the recovery times are longer, since chemical desorption is a slow and irreversible process. SWCNTs with 18% amino groups showed to be selective to formaldehyde when tested against interferences like acetone, CO2, ammonia, methanol and ethanol.

Silicon (Si) nitrogen (N), and phosphorous-nitrogen (P-N) were used to modify MWCNTs and study their gas sensing properties for hydrogen peroxide, sodium hypochlorite and1, 2-dichloromethane, nitrogen, and ammonia [55]. Samples of Si-MWCNTs, N-SWCNTs, and P-N-SWCNTs were prepared by aerosol chemical vapor deposition. It is known that the incorporation of heteroatoms in the CNT structure changes its morphology and thus the reactivity. To evaluate the gas sensing properties of the prepared materials, changes in resistance as function of time were recorded when they were exposed to the different gases. Exposure to N_2 caused the removal of physisorbed water molecules and thus a decrease in the resistance values. Sodium hypochlorite and dichloroethane caused decrease in resistance of pristine MWCNTs and Si-MWCNTs due to charge transfer (electrons) from CNTs to Chlorine atoms and increase in resistance of N-P MWCNTs. Ammonia showed the opposite effect in resistance. These results demonstrate the p-type semiconductor behavior for pristine MWCNTs and Si-MWCNTs and n-type of N-MWCNTs and N-P-MWCNTs. All sensors recovered in 10 min for all gases with the exception of ammonia that exceeded 1 hour.

In an effort to enhance the selectivity of SWCNTs-based vapor sensors, Battie *et al.* worked in the preparation of SWCNTs covered with a mesoporous silica film [56]. Sensors were fabricated by covering a SWCNTs film with a mesoporous silica film via by sol-gel deposition technique. Characterization of the sensors was done by measuring changes in resistance when exposed to 200 ppm of NO_2, NH_3, and H_2O in dry air. A sensor of SWCNTs without the mesoporous silica film prepared and tested under the same conditions. While the SWCNTs sensor showed to be sensitive to the three gases, the sensor based on SWCNTs film covered with mesoporous silica film showed to insensitivity to H_2O, and its sensitivity for NH_3 was considerably reduced. These observations can be explained considering the polarization capabilities and dipole moments of the silanol groups contained in the mesoporous silica layer and the gas molecules. The silanol groups allow the mesoporous silica film to act as a diffusion barrier and allow the physical interaction and entrapment of highly polarized molecules like H_2O and NH_3, avoiding them to get in contact with the SWCNTs layer. On the other hand, the sensitivity to NO_2 was greatly enhanced, compared to the SWCNTs sensor. Compared to H_2O and NH_3, NO_2 has a weaker dipole moment and its diffusion thru the mesoporous silica gel and to the SWCNTs film results easier and thus its enhanced and selective detection.

Computational studies based on SWCNTs doped with heteroatoms have been also reported. *Ab initio* (ABINIT) simulations of CNTs doped with heteroatoms like boron, oxygen and nitrogen were performed to predict the behavior of the doped CNTs and to study their application as gas sensors for Cl_2, CO, NO and H_2 [57, 58]. Density functional theory (DFT) applied in the ABINIT code and the Generalized Gradient Approximation (GGA) were used to perform the calculations. The calculations demonstrated that doping the CNTs with B, O, and N causes a shift in the conduction band of the CNTs. For B and O, the conduction band shifts downward and creates a p-type semiconducting material. On the other hand, N dopant causes the conduction band to shift upward and create an n-type semiconducting material. Calculations also demonstrated that Cl_2, NO, $H_2$and CO considerably affects the NTs density of states (DOS) and Fermi level as the gases become close to their surface. B-doped CNTs can detect CO, NO and H_2 gas molecules, O-doped can detect H_2, Cl_2 and CO and N-doped can detect CO, NO and Cl_2.

Similarly, Hamadanian *et al.* presented a computational study of Al-substituted SWCNTs (10, 0) (2.5% and 25%) and their use as CO gas sensor [59]. DFT calculations (local density approximation with ultrasoft pseudopotential) were used to study the electronic properties of Al-substituted SWCNTs and how those properties are affected by the adsorption of CO molecules. Substitution of one carbon atom with an Al atom causes deformation of the 6-membered ring and increasing the bond length. Doping with Al also alters the DOS and band structure of the CNTs. Since Al has one electron less than C in the valence shell, introduces one electron holes in the band structure, therefore the tube is changed to p-type semiconductor. Calculations showed low adsorption energy for CO on pristine CNTs and that CO does not cause significant changes in the electronic band structure and DOS when adsorbed on pristine CNTs. These results confirm that pristine CNTs are insensitive to CO as result of their weak physical interaction. When CO is adsorbed on both 2.5% and 25% Al-substituted SWCNTs, it causes severe changes in the Band structure near Fermi level. Those changes strongly depended on the site of CNTs and the direction in which the CO molecule interacts. For instance, the most stable adsorption structure is when the C of the CO interacts with the middle point of a C-C bond of CNTs. Even when the adsorption energies of CO in 25% Al-substituted SWCNTs were higher than in 5% Al-substituted SWCNTs, the fact that the conductivity of the proposed material changes, makes them suitable for their use as CO gas sensing material.

Metal oxide	Nanotube Type	Op T. °C	Target (gas)	DL	Response Time	Ref.
ZnO SnO$_2$ TiO$_2$	MWNTs	RT	Ethanol	100ppm*	NS	[60]
Co$_3$O$_4$	SWNTs	RT, 250	NO$_x$	20ppm	NS	[61]
			H$_2$	4%		
ZnO	Pd-COOH SWCNTs	RT	NH$_3$	50ppm*	NS	[62]
	F-SWCNTs					
	N-SWCNTs					
SnO$_2$	SWNTs	200	NO$_2$	2ppm	NS	[63]
SnO$_2$	MWNTs	RT, 150	NO$_2$	1ppm	3min (150C) 4min (RT)	[64]
			CO	2ppm	5min	
WO$_3$	MWNTs	350	H$_2$	100ppm*	Ns	[65]
SnO$_2$	MWNTs	320	Ethanol			[66]
			LPG		21s	
SnO$_2$	O-doped	RT	NO$_2$	100ppb	7min	[67]
	N-doped				1min	
	B-doped				1min	

*Lowest tested concentration

Table 1. Summary of metal oxide NPs used to modify CNTs for gas sensor applications.

4. Conducting polymers and CNT composites

Conducting polymers have been widely used to enhance the sensing properties of CNTs-based sensors. CNTs unique characteristics combined with polymer's delocalized bonds, high permeability and low density have demonstrated that it is possible to detect many different gases with high sensitivity, fast response and good reproducibility. Previous reports on polymer/CNTs-based sensor have been summarized and reviewed [7-9, 12]. However, some challenges to overcome are aggregation or agglomeration of CNTs, thermal stability and selectivity, among others. Polymer/CNTs composites used in resistors, SAW and QMB type of sensors are discussed.

CNTs were used to improve Polyaniline (PANi) poor thermal stability (Figure 4) [68]. The proposed solution to this was the uniform incorporation of CNTs in the polymer network.

Considering that one of the problems of MWCNTs is their aggregation and agglomeration, they were oxyfluorinated under different conditions in order to obtain a better dispersion in aqueous solution and it was found that the better dispersion was obtained with the MWCNTs with the highest oxygen content. The oxyfluorinated MWCNTs were then mixed with aniline and ammonium persulfate (APS) and other chemicals, for in-situ polymerization. Changes in resistance as function of time were used to characterize and evaluate the resulting PANi/MWCNTs composite was for the detection of ammonia (NH_3) in a concentration range of 1-50 ppm. PANi/MWCNTs composite with the highest oxygen content had a uniform composition, improved thermal stability and highest and faster response for 50 ppm of NH_3. The composite was able to detect 1 ppm and showed excellent repeatability for cycling exposures to 50 ppm and it needed heat treatment to accelerate the NH_3 desorption and thus the recovery of the sensor. A possible drawback is the selectivity to NH_3 among gases that can extract protons (H^+) from PANi.

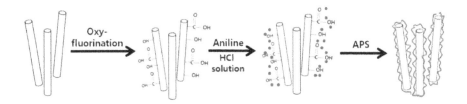

Figure 4. Illustration of the steps to obtain oxyfluorinated CNTs modified with PANI. (From Yun et al. [68]. Copyright © 2012, with permission from Elsevier.)

Mangu *et al.* also worked in the preparation of PANi-MCNTs as well as Poly (3, 4-ethylenedioxythiophene) poly (styrenesulfonate) (PEDOT:PSS)-MWCNTs composites for the detection of 100 ppm of NO_2 and NH_3 [69]. This group studied the effect of dissolving the polymers in different solvents for the gas sensing properties. PANi was dissolved in dimethyl sulfoxide (DMSO), *N, N*-dimethyl formamide (DMF), ethylene glycol (EG) and 2-propanol. PEDOT: PSS was dissolved in DMSO, DMF and 0.1M sodium hydroxide (NaOH). Each polymer solution was spin-coated in plasma treated MWCNTs and evaluated as sensing material. All PANi-MWCNTs composited showed an increase in resistance for NH_3 and a decrease in resistance for NO_2, which is typical of p-type semiconducting composite films. All PANi-MWCNTs composites were selective to NO_2. However, better sensitivities were achieved when PANi was dissolved in 2-propanol and DMSO for NH_3 and for NO_2, respectively. On the other hand, all PEDOT:PSS-MWCNTs composites were also excellent for the detection if both NO_2 and NH_3. PEDOT: PSS-MWCNTs (prepared without any solvent) showed to be more sensitive to NH_3 and PEDOT:PSS dissolved in NaOH to NO_2.

Sayago, *et al.* have worked on the preparation of different composites using polymers with small percentages of CNTs as sensitive layers for surface acoustic wave (SAW) gas sensors [70-72]. Composites of polyisobutilene (PIB), polyepichlorohydrin (PECH) and polyetherurethane (PEUT) with 2% and 5% of MWCNTs were prepared and used to detect volatile organic compounds (VOCs) at room temperature using SAW-sensor arrays. All the samples showed responses (frequency shifts) for octane and toluene (25-200 ppm), even though some samples were more sensitive than the others. For instance, samples with PIB/5%MWCNTs showed higher sensitivity for octane while PECH/2%MWCNTs and PEUT/2%MWCNTs were more sensitive to toluene. The difference in sensitivity is attributed to the difference in affinity between polymers and VOCs due to their respective polarities. The detection and recovery times were fast and fully reversible, which means that the main interaction is physisorption. The role of the MWCNTs is unknown. In general, their presence in the composite showed to improve sensitivity and reduce the limits of detection (LOD) but did not affect selectivity, response and recovery times.

Another SAW gas sensor was reported by Viespe et al. for the detection of methanol, ethanol, toluene using different polyethyleneimine (PEI)-based nanocomposite as sensitive layer, including MWCNTs-PEI [73]. In general, frequency shifts were proportional to the gas concentration and MWCNTs-PEI sensors showed better response time and higher sensitivity than the PEI-sensor. However, it did not show the best LOD when compared to the other PEI-based nanocomposite. The MWCNTs-PEI sensors showed higher sensitivity towards toluene and lower sensitivity to methanol when compared to ethanol.

Biopolymer/CNTs composites for chemical vapor sensors were produced by using two different biopolymers, cellulose, the most naturally abundant one and poly (lactic acid) (PLA). Considering that previous studies have shown that a homogeneous distribution of MCNTs in the cellulose matrix can improve the polymer's mechanical and electrical characteristics, MWCNTs were functionalized with imidazolide groups and covalently attached it to cellulose chains [74]. The resulting material, a paper-like film, was then used as sensing element for the detection of methanol, ethanol, 1-propanol and 1-butanol at ppm levels. Responses were measured by changes in resistance and were found to be reversible and consistent for all the tested vapors. However, the sensor only showed linear responses as function of concentration for 1-propanol in the range of 400-3600 ppm. The other composite, PLA/MWCNTs was prepared by doping the biopolymer with2 and 5% of MWCNTs and annealing, in order to understand the effect of MWCNTs in the crystallinity of the polymer and its performance in the detection of toluene, water, methanol and chloroform [75]. PLA/2%-MWCNTs showed highest responses for all gases, when compared to PLA/3%-MWCNTs. However, it was found that all samples were selective to chloroform. Moreover, annealing the samples showed a decrease in the responses that were significantly lower than the untreated ones. Annealing did not affect the selectivity to chloroform but it considerably affected its sensitivity.

Plasma-treated MWCNTs (p-MWCNTs) polyimide (PI) composite films (p-MWCNTs-PI) were developed by Yoo et al. in an effort to overcome some of the problems presented by PI-based resistive-type sensors, e. g. nonlinear sensitivity to relative humidity [33]. When tested as sensing material for 10-95% relative humidity (RH), p-MWCNTs-PI showed better sensitivity and linear response (resistance) as function of humidity, when compared to pristine-MWCNT-PI and pristine PI. The increase of p-MWCNTs content inp-MWCNTs-PI films not only showed improved the linearity and sensitivity but also lower resistance values. Lower resistance values might improve the performance at low RH range, a problem that PI resistive-type sensors present because of their high resistance values.

Yuana and coworkers prepared a sensor array based on polymer/MWCNTs composites for the selective detection of chloroform ($CHCl_3$), tetrahydrofuran (THF) and methanol (MeOH) [76]. Ethyl cellulose (EC), poly [methyl vinyl ether-alt-maleic acid] (PMVEMA), hydroxypropyl methyl cellulose (HPMC), poly (alpha-methylstyrene) (PMS), poly (vinyl benzyl chloride) (PVBC) and poly (ethyleneadipate) (PEA) were the polymers used to prepare the polymer/MWCNTs composites and to provide uniqueness to each sensor in terms of their physical and chemical characteristics like molecular structure, polymer length, polarity and intermolecular forces. Changes in resistance as function of time were recorded for the sensor array when exposed to the different gases at different temperatures (30 40, 50, 60 °C) and 50-60% R. H. The sensors showed to be selective to the three gases at the different temperatures when in presence of vapor molecules of chloride, cyclic oxide and hydroxide groups. The decreasing order of sensitivity concurred with the order of decreasing conductivity: PEA/MWCNTs >EC/MWCNTs >PMVEMA/MWCNTs >PVBC/MWCNTs >HPMC/MWCNTs >PMS/MWCNTs.

SWCNTs modified with Poly- (D) glucosamine (Chitosan) (SWCNTs-CHIT) were as high performance hydrogen sensor [77]. Three types of sensors were prepared: SWCNTs deposited on glass substrate (type 1), SWCNTs deposited over a glass substrate modified with CHIT-film (type 2), Chit-film deposited over SWCNTs deposited over a glass substrate. Each type of sensor showed a different changes in resistance when exposed to 4% H_2 in air. Increase in resistance was observed for three types of sensors and good recovery but for type (3) sensor. Sensors modified with CHIT showed better sensitivity. Moreover, the authors explained that the improved sensitivity was far higher than that reported for Pd-SWCNTs based sensors, which are commonly used for H_2 detection. The enhanced performance of SWCNTs-CHIT can be explained by the strong interaction/ binding of the H_2 molecules with/to the –OH and $–NH_3$ groups contained in CHIT, which also explains the poor recovery of type 3 sensor.

SWCNTs were modified with Polypyrrole (PPy) and 5, 10, 15, 20-tetraphenylporphyrine (TPP) to prepare composites for the detection of 1-butanol in nitrogen using a quartz microbalance (QMB) [78]. The QMB gold electrodes were coated with PPy/SWCNTs-COOH, PPy/SWCNTs-COOH/SWCNTs-TPP via electropolymerization. Frequency shifts of the quartz crystal resonator as function of time were measured for the sensors when exposed to 1-butanol in ppb concentration range. Even though both composites showed good and higher response magnitudes than QMB prepared with other composites that did not contain CNTs, PPy/SWCNTs-COOH/SWCNTs-TPP showed better performance than PPy/SWCNTs-COOH. The results demonstrate that the incorporation of CNTs enhanced the sensitivity towards the detection of 1-butanol.

Lu *et al.* reported a sensor array containing pristine SWCNTs, Rh-loaded SWCNTs, PEI/ SWCNTs and other CNTs with different coatings and loadings for the detection of hydrogen peroxide (H_2O_2) [79]. The measurements of changes in resistance as function of time were used to analyze the sensor array performance when exposed to H_2O_2. Pristine SWCNTs showed strong increases in resistance and fast responses for H_2O_2 and an estimated DL (by IUPAC definition) of 25 ppm. But when the sensor array was exposed to H_2O and CH_3OH in order to test its selectivity, pristine SWCNTs showed also good responses for both chemicals, which means that the discrimination capabilities towards H_2O_2 are limited. On the other hand, the PEI/SWCNTs sensors were sensitive to H_2O_2, showing decreases in resistance for each exposure. The PEI/SWCNTs did not show significant changes in resistance when exposed to H_2O and CH_3OH, which makes it selective to H_2O_2 under the tested conditions.

Polymer	CNT Type	Sensor Configuration	Target	DL	Ref.
PECH, PEUT, PIB	MWCNTs	SAW	Toluene	1.7-12.2ppm	[70 72]
			Octane	9.2-12.7ppm	
				NS	
Cellulose	MWCNTs	Resistor	Methanol	650ppm*	[74]
			Ethanol	672ppm*	
			1-propanol	635ppm*	
			1-butanol	687ppm*	
PANi	MWCNTs	Resistor	NH_3	1ppm	[68]
PEDOT:PSS	MWCNTs	Resistor	NH_3	100ppm*	[69]
			NO_2		
PI	MWCNTs	Resistor	Humidity	10%*	[33]
PEI	MWCNTs	SAW	Ethanol	176.5	[73]
			Methanol	184.2	
			Toluene	170.6	
PLA	MWCNTs	Resistor	VOC	NS	[75]
CHI	SWCNTs	Resistor	H_2	4%*	[77]
EC	MWCNTs	Resistor	THF, CH_3Cl_2, MeOH	NS	[76]
PMVEMA					
HPMC					
PMS					

PVBC					
PEA					
PPy	SWCNTs	QMB	1-butanol	46ppb*	[78]
PPT					
PEI	SWCNTs	resistor	H_2O_2	NS	[79 80]

* Lowest detected concentration

Table 2. Summary of polymers used for the preparation of polymer/CNTs-based sensors.

5. Metal nanoparticlesdecorated CNTs

Electronic, physical and chemical properties of metallic nanoclusters are usually sensitive to the changes in environment [81]. CNTs decorated with metallic nanoparticles (NPs) have been widely used to achieve selectivity and improve the sensitivity, response time and DLs for a variety of gas detections. Layer by layer, electrodeposition, chemical deposition, electrochemical deposition and sputtering are the methods used to prepare the metallic NP-CNTs composites discussed in this section.

SWCNTs films were modified with Pd NPs using sputtering method [82]. After apply different deposition times (40 – 160 s), it was found that 120 s was the optimum deposition time to obtain enhanced sensor response for 1% H_2 in dry air at 50 °C. A typical response curve (i_{gas}/i_{air} vs. time) showed differences in response and recovery between the first and following H_2 sensing cycles. FTIR studies were used to support and explain those differences and mechanisms of detection. The first cycle showed an overall larger electrical current in the presence of $H_2$and then it reached a new steady state. When the atmosphere was change to dry air, the current did not go to its original value but remained in the steady state, which is considered as an irreversible response. The explanation to this is that is atomized by the Pd NPs and spilled to the surface of the MWCNTs, occurring the chemical and irreversible reaction of hydrogenation of the carbonyl groups of the MWCNTs at the first cycle. The second cycle and following ones started at the steady state where the first cycle finished and the electrical current showed a decrease in the presence of H_2 and when the atmosphere was changed to dry air, the electrical current recovered back to where the cycle started. This reversible behavior is explained as physisorption of H_2 molecules onto Pd/SWCNTs.

Pd/MWCNTs and Pt-Pd/MWCNTs composites were tested for the detection of H_2 in a concentration range of 20 ppm– 2% in N_2 and 200 ppm – 2% in air [83]. Composites were prepared by growing CNTs yarns and then covered them with a layer of Pd NPs or sequentially deposited layers of Pd and Pt NPs, using a recently developed technique called self-fuelled electrodeposition (SFED). Exposure to 1% H_2using N_2 with 1% air as carrier gas. As with other Pd/CNTs-based sensors [82], an initial irreversible drop in resistance was observed, and after that, the sensor reached a steady state. A stable baseline was established just after a couple of exposure/recovery cycles. Pd-MWCNTs was not able to detect H_2 concentrations below 20 ppm but with the Pt-Pd/MWCNTs composites it was possible to detect concentrations as small as 5 ppm (0.0005%). The sensor saturated at 100 ppm and higher concentrations. When the same experiments were performed using air as carrier gas, it was found that the detection limit for both composites decreased.

A Pd/MWCNTs flexible substrate H_2 sensor was fabricated using the layer-by-layer technique [84]. For this, an Au IDE was sputtered over a polyester (PET) film, followed by the fabrication of a poly (4-styrenesulfonic acid-co-maleic acid)/poly (allylamine hydrochloride) (PSSMA/PAH) bilayer film. Then, a MWCNTs layer was deposited over the PSSMA/PAH bilayer film and decorated with Pd NPs using chemical deposition. The sensors were exposed to H_2 in a range of concentrations between 200 and 40000 ppm at room temperature and 53% RH. The rapid response and higher sensitivity of the Pd/MWCNTs when compared to plain MWCNTs is attributed to the well known catalytic effect of Pd NPs. Because the detection mechanism is the dissociation of the H_2 molecules on the surface of the Pd NPs, the linear relation was found to be between the sensor's response and the square root of the concentration. It also showed to be selective to H_2 at concentrations higher than 1000 ppm (vs. NH_3, CO, CH_4 and others), to be highly reproducible, to have long-term stability and comparable to similar sensors fabricated in rigid substrates.

Zilli et al. demonstrated that the H_2sensing capacity of the Pd/MWCNTs nanocomposite is affected by the different stages of purification of the CNTs [85]. Pd NPs were chemically deposited on pristine MWCNTs (Pd-CNT-P), gas-phase oxidized MWCNTs (Pd-CNT-O) and gas-phase oxidized/acid treated MWCNTs (Pd-CNT-A) and used as H_2 sensing material. All three nanocomposites showed increase in resistance as function of time in the presence of $500\mu L$ (STP) of H_2 and an immediate decrease when the H_2 disappeared from the testing chamber. However, Pd-CNT-O showed to be more sensitive than Pd-CNT-P and Pd-CNT-A, which both showed similar responses. The reason for these results is that for CNT-P, catalytic Fe NPs, used for the CNT growth, are encapsulated inside the CNTs and in CNT-A, the Fe NPs were removed the during the acid treatment. After the gas-phase oxidation process (CNT-O), the Fe NPs become exposed and at the same time, oxygen-containing groups are formed in the surface of CNTs, which act as additional anchoring site for Pd NPs and thus making the Pd-CNT-O sample more sensitive to H_2.

Pd/SWCNT composites were prepared by using a poly (amido amine) (PAMAM) dendrimer assisted synthesis, followed by a pyrolysis step to remove the dendrimers [86]. For H_2 sensing experiments, the samples were deposited over Ti/Au electrodes and changes in resistance as function of time were measured. The Pd/SWCNTs samples showed to be more sensitive to 10, 000 ppm of H_2 at room temperature, when compared to samples of chemically reduced Pd on SWCNTs (without the presence of dendrimer) and Pd/PAMAM-SWCNTs. Moreover, these samples were able to detect all of the concentrations in a 10-100 ppm range. It was concluded that not only the dendrimers provided more nucleation sites for the Pd NPs and thus higher NPs density, but also that the removal of the dendrimers thru the pyrolysis step reduces the distance between the NPs and the SWCNTs and consequently reduces the delay in electron transfer, allowing faster response times.

Double wall carbon nanotubes (DWCNTs) have shown to have longer length, compared to SWNTs, provide a better percolation behavior, the possibility of modifying the outer layer without modifying the inner one and flexibility are some of their attractive characteristics. Considering those characteristics, Rumiche *et al.* evaluated Pd NPs/ (DWCNTs) composites as room temperature H_2 sensor (Figure 5) [87]. Different amounts of DWNTs (15 and 20 μL) were deposited over silicon oxide substrates and decorated with 1, 3 and 6nm layers of Pd NPs. To evaluate their sensing properties, changes in resistance as function of time were recorded when exposed to 3%, 2%, 1%, 0.5%, 0.3%, 0.2%, 0.1%, and 0.05% of H_2 in dry air. Samples containing 1nm Pd layer were unsuccessful detecting any of the tested concentrations. On the other hand, samples containing 3 and 6nm Pd layer showed overall similar performance in terms of increases of resistance and recovery. When analyzing the results for the lowest tested H_2 concentration (0.05%)it was found that the increase in resistance was comparable for samples with same Pd layer thickness and different DWCNTs content (e. g. 3nm thick Pd layer deposited over 15μL DWCNTs was comparable to of 3nm thick Pd layer over 20μL DWCNTs). However, the increase in Pd coating thickness produced a reduction in the response. The obtained results confirmed that the combination of the amount of DWCNTs and the Pd-layer thickness directly affects the sensitivity of the sensors.

Another H_2 sensor based on N-doped MWCNTs electrochemically decorated with Au NPs (Au-NMWCNTs) was presented by Sadek and collaborators [88]. N-doped MWCNTs were chosen because they have enhanced surface reactivity and chemically active sites for the nucleation of Au NPs during the electrodeposition process. NPs of different sizes were obtained with variation of electrodeposition potential. Changes in resistance as function of time were used to evaluate the performance of the Au-NMWCNTs sensors when exposed to different H_2 concentration between 0.06% and 1%. Sensitivity, response time and recovery time highly depended on the size of the AuNPs: the smallest the size the better the sensitivity and the shorter the response and recovery times.

Penza *et al.* worked in the modification of SWCNTs with Pt, Ru and NiNPs to monitor toxic, landfill, and greenhouse gases (CO_2, CH_4, NH_3, and NO_2) [89, 90]. Pt NP layers with thickness of 8, 15 and 30 nm were sputtered over SWCNTs films and exposed to the different gases at an operation temperature of 120°C. Changes in resistance as function of time showed that Pt-SWCNTs had better sensitivity than unmodified SWCNTs. It was also found that the sensitivity depended on the layer thickness. For instance, 8 nm Pt layer-SWCNTs showed highest sensitivity for NO_2 and CH_4 and 15 nm Pt layer-SWCNTs showed better sensitivity for NH_3 and CO_2. In a similar study, a sensor array containing Pt, Ru, and Ag NPs sputtered over SWCNTs with a thickness of 5nmwas able to detect and selectively discriminate between landfill gases. Concentrations as low as 100 ppb of NO_2 were selectively detected at a temperature of 120 °C.

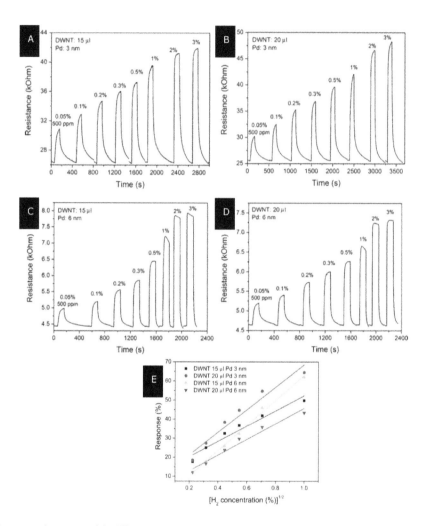

Figure 5. A-D) Responses of the different Pd/DWCNTs composites when exposed to H$_2$. (E) Calibration curves for the samples presented in figures (A-D). (From Rumiche *et al.* [87] Copyright © 2012, with permission from Elsevier.)

Lu *et al.* used pristine SWCNTs, fluorinated SWCNTs (F-SWCNTs) and rhodium doped SWCNTs (Rh-SWCNTs) and other various coatings and dopings on the SWCNTs for the room temperature detection of formaldehyde (HCOH) [80]. The measurements of changes in resistance as function of time when the array was exposed to the different concentrations of formaldehyde were used to analyze the sensor array performance. When exposed to 0.71 ppm formaldehyde, pristine SWCNTs showed the higher response, followed by Rh/SWCNTs and F-SWCNTs, which were less sensitive. The other CNTs with different coating and loadings were insensitive to formaldehyde. However, when the array was exposed to formaldehyde at concentrations as low as 0.01 ppm (10ppb), Rh/SWCNTs sensor showed to be more sensitive and the presented an estimated DL (by IUPAC definition) of 10ppb. The DL for pristine SWCNTs and F-SWCNTs sensors were 15ppb and 20ppb, respectively. The three sensors presented very fast response (~18sec) and recovery time of approximately 1 minute.

Theoretical studies have been used to study the interaction between Pd and Pt NPs decoratedCNTs and a wide variety of gases for different applications. Zhou *et al.* used the density functional theory (DFT) to study the adsorption and interaction of SO_2, CH_3OH, and CH_4 with Pd-SWCNTs [91]. The replacement of a central C atom of the CNTs with a Pd atom causes structural deformations. As SO_2 is adsorbed, there is a charge transfer from Pd-SWCNT to SO_2. As for CH_3OH, the appropriate adsorption conformation is thru the lone par of the oxygen of CH_3OH and thus occurring an overall charge transfer from CH_3OH to Pd-SWCNT. The interaction between Pd-SWCNT and CH_4 is similar in that the charge transfer occurs from Pd-SWCNT to the gas. However, the interaction between Pd-SWCNTs and CH_4 is not as strong as the interaction between Pd-SWCNT and SO_2 or even CH_3OH.

NP	NT Type	Target	Sensor Configuration	DL	Ref.
Pd	MWCNT	H_2	Resistor	1%	[82]
Pd	MWCNTs	H_2	Resistor	10000ppm*	[84]
Pd	MWCNTs	H_2	Resistor	NS	[85]
Pd-Pt	MWCNTs	H_2	Resistor	5ppm	[83]
Pd	SWCNTs	H_2	Resistor	10ppm	[86]
Pd	DWCNTs	H_2	Resistor	0.05	[87]
Pt, Ru, Ag	SWCNTs	$CO_2 CH_4$, NH_3, NO_2	Resistor	100ppb NO_2	[89 , 90]
Au	N-doped MWCNTs	H_2	Resistor	0.06%*	[88]
Rh	SWCNTs	HCOH	Resistor	10ppb	[80]

*Tested concentration

^For concentrations >1%

Table 3. Metallic nanoparticles used to decorate SWCNTs for gas sensing applications

Li and co workers investigated the adsorption of CO and NO on SWCNTs-decorated with Pd and Pt using first-principle calculations [92]. It was found that the electronic properties of SWCNTs change upon modification with Pd or Pd atoms. The semi-conductive band gap is decreased compared to pristine SWCNTs. The reason for the observed decrease in the band gap is due to charge transfer from the Pd and Pt atoms to the surface of the SWCNTs. Different from pristine SWCNT that show poor adsorption, Pd-SWCNTs and Pt-SWCNTs showed to chemisorb CO molecules as well as NO. However, Pt-SWCNTs showed bigger binding energy and charge transfer than Pd-SWCNTs. The formation of C-Pd, N-Pd, C-Pt, and N-Pt bonds demonstrate that the metal atoms provide additional adsorptions sites for gases and open the possibility to use both materials as sensors for the detection of CO and NO.

6. Nanostructured oxides mixed with CNTs

Sensors made of metal oxides films have been used for a long time because of they provide high sensitivity for the detection of a wide variety of gases. However, their major drawback is their elevated operating temperatures. The development of metal-oxide NPs based films and nanocomposites has shown advantages like higher surface area and porosity, high catalytic activity, efficient charge transfer and adsorption capacity. However, it has been demonstrated that the improvements in gas detection at low temperature for CNTs/MO-based sensors is due to the introduction of CNTs in the nanocomposite.

Tin oxide (SnO_2)/MWCNTS were synthesized using different ratios of tin dioxide precursor and plasma treated MWCNTs (Figure 6) [64]. The composites was tested for 2, 10, 20 ppm for CO and 50 100, 500, 1000 ppb of NO_2 in dry air at both room temperature and 150 °C. Pure tin oxide films and pure plasma treated MWCNT were also tested for comparison purposes. Pure tin oxide films were unresponsive to all of the tested concentrations at the two different temperatures because both room temperature and 150 °C are too low when compared to the operation temperature for pure tin oxide-based sensors. On the other hand, pure plasma treated MWCNT responded to both gases at room temperature but not at 150 °C. As for the SnO_2/MWCNTS composite, higher sensor response for both gases was achieved from samples prepared with an intermediate ratio of tin dioxide precursor and plasma treated MWCNTs (i. e. 20mL and 12mg, respectively), especially when operated at room temperature. Response time for 1 ppm of NO_2 was 3minutesat 150 °C and 4 minutes at room temperature. Response time for 2 ppm of CO is stated to be 5 minutes but the temperature was not specified. SnO_2/MWCNTS showed higher sensor response to NO_2 than to CO, and was also sensitive to humidity changes.

Different composite synthesis temperature can affect the sensor performance. SnO_2/SWCNTs composites were synthesized at different oxidizing temperatures (300-600 °C)for testing the effect of temperature in their morphology, structure and gas sensing properties in the detection of NO_X [63]. The synthesized composites were exposed to 60 ppm of NO_X at 200 °C and it was found that the ones synthesized at 400 °C showed higher response. From here, composites synthesized at 400 °C were exposed to 30 ppm NO_X at different operating temperatures and it was determined that the optimum operation temperature was 200 °C. Concluding that the optimum oxidizing and operating temperatures were 200 °C and 400 °C, respectively, the samples were then exposed to different concentrations of NO_X. Under the aforementioned conditions, the SnO_2/SWCNTs composite showed improved performance for the detection of NO_X when compared to thin films of SWCNTs or SnO_2.

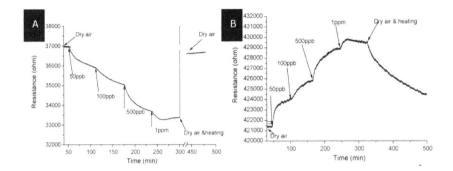

Figure 6. Response of SnO$_2$/MWCNTs to NO$_2$ at (A) room temperature and (B) 150 °C. (From Leghrib *et al.* [64]. Copyright © 2010, with permission from Elsevier.)

Wongchoosuk, *et al.* were the first to report the preparation of MWCNT-doped tungsten oxide (WO$_3$) thin films for H$_2$ sensing application [65]. The thin films of MWCNT-doped WO$_3$ and undoped WO$_3$ (for comparison purposes) were prepared using the electron beam (e-beam) evaporation technique and exposed to 1000 ppm of H$_2$ at different temperatures (200-400 °C). It was determined that 350 °C was the optimum operation temperature. Overall, MWCNT-doped WO$_3$ thin films showed higher responses for H$_2$ at any operating temperature when compared to the undoped WO$_3$ thin film. To demonstrate selectivity, the response of the MWCNT-doped WO$_3$ thin films was measured in presence of H$_2$, ethanol (C$_2$H$_5$OH), methane (CH$_4$), acetylene (C$_2$H$_2$) and ethylene (C$_2$H$_4$) at ppm level concentrations and operating temperature of 350 °C. It was concluded that the MWCNT-doped WO$_3$ thin filmswere selective to H$_2$ because they showed stronger response for H$_2$, much weaker responses for C$_2$H$_5$OH, CH$_4$, C$_2$H$_2$ and insensitivity ethylene (C$_2$H$_4$).

MWCNTs treated with nitric acid were used to fabricate MWCNTs-doped SnO$_2$sensors for the detection of ethanol and liquid petroleum gases (LPG) [66]. Sensors were tested to 100-1000 ppm of ethanol and 1000-10, 000 ppm of LPG at different operation temperatures in the range of 10-360 °C. The detection of both chemicals was improved when the operating temperature was 350 °C or lower and it was determined that the optimum operating temperature is 320 °C. The MCNTs-doped SnO$_2$ composite showed better selectivity for LGP than for ethanol and the calibration curve showed the sensors saturated at concentrations higher than 5000 ppm. The90% response and recovery time were 21s and 36s, respectively. When undoped SnO$_2$ sensors were exposed to 250 ppm of ethanol and 2500 ppm of LPG, they showed higher sensitivity to ethanol than to LPG at operation temperature range of 190 - 360 °C. Considering the obtained results, the selectivity of the MCNTs-doped SnO$_2$ composite for LGP can be attributed to the presence of MWCNTs but further studies are required.

Nanocomposite structures of cobalt oxide (Co$_3$O$_4$) and SWCNTs were prepared using a polymer assisted deposition (PAD) method [61]. For this, polyethyleneimine (PEI) was the polymer used to bind the cobalt ions from and adjust the viscosity of the solution during the deposition process in order to get a homogeneous distribution of the particles on the

SWCNTs thin film. The Co_3O_4/SWCNTs composite sensor was tested for the detection of NO_x in a concentration range of 20-100 ppm at room temperature. It showed proportional increases in response as function of concentration, poor recovery at room temperature and good recovery at 250 °C. Higher responses of the Co_3O_4/SWCNTs composite when compared to pristine CNTs are attributed to the high adsorption power of Co_3O_4 particles. The composite was also exposed to 4% of H_2 in air, and showed enhanced responses than pure SWCNTs at room temperature and than Co_3O_4 films at both room temperature and 250 °C.

MO NPs (ZnO, SnO_2, TiO_2) and MWCNTs composites were simultaneously grown on silicon and silica on silicon substrates by catalytic pyrolysis method and used for gas sensing [60]. Current differential-voltage (ΔI-V) curves were recorded for all the prepared composites, while to 100 ppm ethanol. TiO_2/MWCNTs showed better sensitivity (defined as ΔI/I-V) when compared to pure MWNT film, ZnO/MWCNTs and SnO_2/MWCNTs.

N-doped, B-doped and O-doped CNTs were used to prepare doped-CNTs/SnO_2hybrids [67]. All doped-CNTs and doped-CNTs/SnO_2hybrids were used to study the effect of functional groups on their gas sensing properties for 100, 200, 500, 1000ppb of NO_2 at room temperature. The responses were as follows: B-doped hybrid> N-doped hybrid>O-doped hybrid. All doped-CNTs/SnO_2hybrids responded better than N-doped and B-doped and O-doped CNTs. B-doped-CNTs/SnO_2 hybrids showed an improvement in the response time when compared to bare CNTs and recovered its baseline, which was not achieved with B-doped CNTs. The high sensitivity and improved performance achieved with the B-doped and N-doped-CNTs/ SnO_2 hybrids for low concentrations of NO_2 at room temperature are attributed to two main factors: the interaction of the N2 gas with the n-SnO_2/p-CNTs heterostructure that affects the conduction of the CNTs and the addition of new functionalities (i. e. B and N atoms) to the CNTs surface that affects the electronic density of states and Fermi level and consequently, its conductivity.

A combination of ZnO layer with functionalized MWCNTs for the room temperature detection of NH_3 has been reported by Tulliani and coworkers [62]. Samples of Pd-doped/COOH-MWCNTs, N-MWCNTs, and F-MWCNTs were deposited over a screen-printed ZnO layer. The materials were evaluated by measuring changes in resistance as the sensors were exposed to NH_3 at room temperature, in a concentration range 0-75 ppm and different relative humidity levels. The sensor based on ZnO with Pd-doped/COOH-MWCNTs was the only one that showed sensitivity to humidity. When exposed to NH_3, all sensors showed a decrease in electrical resistance but did not show better DL than other graphite-based sensors prepared under the same conditions.

7. Conclusion

Modification and functionalization of CNTs have shown to greatly improve the sensitivity and selectivity of CNTs-based sensors. For instance, great improvements for room temperature detection of different gases have been reported, especially when using metal oxide/ SWCNTs composites. Another subject of high interest is the development of deposition methods and synthesis of Pd/CNTs for H_2 detection at room temperature. Interestingly, there is an increase in the tendency of combining other materials with modified CNTs. For example, CNTs decorated with metal NPs embedded in a polymer matrix or CNTs doped or CNTs doped with heteroatoms and decorated with NPs or metal oxides are some composites that have been successfully used as gas sensing materials. But not only the characteristics of the CNTs have contributed to these improvements. In fact, the reported improvements are attributed to the combination of materials and the intrinsic characteristics of the composites. This trend of combining materials demonstrates that the there is broad range of possibilities for the design of new materials to meet the requirements of an ideal sensor by showing selectivity for different gases, sensitivity at low concentrations, fast response, and room temperature operation among others.

Acknowledgements

The authors would like to acknowledge the NASA-URC Center for Advanced Nanoscale Materials (CANM) Grant # NNX08BA48A and NNX10AQ17A, NASA GSRP fellowship under Grant # NNX09AM23H and NASA Ames Research Center-Nanosensor Group

Author details

Enid Contés-de Jesús[1], Jing Li[2] and Carlos R. Cabrera[1*]

1 Department of Chemistry and NASA-URC Center for Advanced Nanoscale Materials, University of Puerto Rico, Puerto Rico

2 NASA-Ames Research Center, Moffett Field, California, USA

References

[1] Niyogi S, Hamon MA, Hu H, Zhao B, Bhowmik P, Sen R, et al. Chemistry of single-walled carbon nanotubes. Acc Chem Res. 2002 Dec;35(12):1105-1113.

[2] Kong J, Franklin NR, Zhou CW, Chapline MG, Peng S, Cho KJ, et al. Nanotube molecular wires as chemical sensors. Science. 2000 Jan;287(5453):622-625.

[3] Ciraci S, Dag S, Yildirim T, Gülseren O, Senger RT. Functionalized Carbon Nanotubes And Device Applications. J Phys: Condens Matter. 2004;16(29):R901-R960.

[4] Chu H, Wei L, Cui R, Wang J, Li Y. Carbon nanotubes combined with inorganic nanomaterials: Preparations and applications. Coord Chem Rev. 2010;254(9–10): 1117-1134.

[5] Liu S, Shen Q, Cao Y, Gan L, Wang Z, Steigerwald ML, et al. Chemical functionaliza-
     tion of single-walled carbon nanotube field-effect transistors as switches and sensors.
     Coord Chem Rev. 2010;254(9):1101-1116.

[6] Varghese OK, Kichambre PD, Gong D, Ong KG, Dickey EC, Grimes CA. Gas sensing
     characteristics of multi-wall carbon nanotubes. Sens Actuators, B. 2001;81(1):32-41.

[7] Wang Y, Yeow JTW. A Review of Carbon Nanotubes-Based Gas Sensors. Journal of
     Sensors. [electronic]. 2009;doi: 10.1155/2009/493904 (accessed 13 April 2012).

[8] Zhang T, Mubeen S, Myung NV, Deshusses MA. Recent progress in carbon nano-
     tube-based   gas   sensors.   Nanotechnology.   2008;doi: 332001
     10.1088/0957-4484/19/33/332001 (accessed 13 April 2012).

[9] Zhang W-D, Zhang W-H. Carbon Nanotubes as Active Components for Gas Sensors.
     Journal of Sensors. 2009;doi: 10.1155/2009/160698 (accessed 13 April 2012).

[10] Di Francia G, Alfano B, La Ferrara V. Conductometric Gas Nanosensors. Journal of
     Sensors. 2009;doi:10.1155/2009/659275.

[11] Andzelm J, Govind N, Maiti A. Nanotube-based gas sensors - Role of structural de-
     fects. Chem Phys Lett. 2006 Apr;421 (1-3):58-62.

[12] Dai LM, Soundarrajan P, Kim T. Sensors and sensor arrays based on conjugated pol-
     ymers and carbon nanotubes. Pure Appl Chem. 2002 Sep;74 (9):1753-1772.

[13] Goldoni A, Larciprete R, Petaccia L, Lizzit S. Single-wall carbon nanotube interaction
     with gases: Sample contaminants and environmental monitoring. JACS. [Article].
     2003 Sep;125 (37):11329-11333.

[14] Li J, Lu YJ, Ye Q, Cinke M, Han J, Meyyappan M. Carbon nanotube sensors for gas
     and organic vapor detection. Nano Lett. 2003 Jul;3 (7):929-933.

[15] Novak JP, Snow ES, Houser EJ, Park D, Stepnowski JL, McGill RA. Nerve agent de-
     tection using networks of single-walled carbon nanotubes. Appl Phys Lett. [Article].
     2003 Nov;83 (19):4026-4028.

[16] Valentini L, Armentano I, Kenny JM, Cantalini C, Lozzi L, Santucci S. Sensors for
     sub-ppm NO2 gas detection based on carbon nanotube thin films. Appl Phys Lett.
     2003 Feb;82 (6):961-963.

[17] Lu YJ, Li J, Han J, Ng HT, Binder C, Partridge C, et al. Room temperature methane
     detection using palladium loaded single-walled carbon nanotube sensors. Chem
     Phys Lett. 2004 Jun;391 (4-6):344-348.

[18] Wang SG, Zhang Q, Yang DJ, Sellin PJ, Zhong GF. Multi-walled carbon nanotube-
     based gas sensors for NH_3 detection. Diamond Relat Mater. 2004 Apr-Aug;13 (4-8):
     1327-1332.

[19] Lee CY, Strano MS. Understanding the Dynamics of Signal Transduction for Adsorption of Gases and Vapors on Carbon Nanotube Sensors. Langmuir. 2005 2005/05/01;21 (11):5192-5196.

[20] Penza M, Antolini F, Vittori-Antisari M. Carbon nanotubes-based surface acoustic waves oscillating sensor for vapour detection. Thin Solid Films. 2005 Jan;472 (1-2): 246-252.

[21] Lu YJ, Partridge C, Meyyappan M, Li J. A carbon nanotube sensor array for sensitive gas discrimination using principal component analysis. J Electroanal Chem. 2006 Aug;593 (1-2):105-110.

[22] Robinson JA, Snow ES, Badescu SC, Reinecke TL, Perkins FK. Role of Defects in Single-Walled Carbon Nanotube Chemical Sensors. Nano Lett. 2006 2006/08/01;6 (8): 1747-1751.

[23] Star A, Joshi V, Skarupo S, Thomas D, Gabriel J. Gas Sensor Array Based on Metal-Decorated Carbon Nanotubes. J Phys Chem B. 2006;110:21014 - 21020.

[24] Trojanowicz M. Analytical applications of carbon nanotubes: a review. TrAC, Trends Anal Chem. 2006 May;25 (5):480-489.

[25] Kuzmych O, Allen B, Star A. Carbon nanotube sensors for exhaled breath components. Nanotechnology. 2007; doi:10.1088/0957-4484/18/37/375502.

[26] Schlecht U, Balasubramanian K, Burghard M, Kern K. Electrochemically decorated carbon nanotubes for hydrogen sensing. Appl Surf Sci. 2007 Aug;253 (20):8394-8397.

[27] Valcarcel M, Cardenas S, Simonet BM. Role Of Carbon Nanotubes In Analytical Science. Anal Chem. 2007 Jul;79 (13):4788-4797.

[28] Endo M, Strano MS, Ajayan PM. Potential applications of carbon nanotubes. Carbon Nanotubes. 2008;111:13-61.

[29] Kauffman DR, Star A. Carbon nanotube gas and vapor sensors. Angewandte Chemie-International Edition. 2008;47 (35):6550-6570.

[30] Li C, Thostenson ET, Chou T-W. Sensors and actuators based on carbon nanotubes and their composites: A review. Compos Sci Technol. 2008;68 (6):1227-1249.

[31] Zanolli Z, Charlier JC. Defective carbon nanotubes for single-molecule sensing. Phys Rev B. 2009;doi:10.1103/PhysRevB.80.155447.

[32] Goldoni A, Petaccia L, Lizzit S, Larciprete R. Sensing gases with carbon nanotubes: a review of the actual situation. J Phys: Condens Matter. 2010; doi: 10.1088/0953-8984/22/1/013001.

[33] Yoo K-P, Lim L-T, Min N-K, Lee MJ, Lee CJ, Park C-W. Novel resistive-type humidity sensor based on multiwall carbon nanotube/polyimide composite films. Sens Actuators, B. 2010;145 (1):120-125.

[34] Lu Y, Meyyappan M, Li J. Fabrication of carbon-nanotube-based sensor array and interference study. Journal of Materials Research. 2011;26 (26):2017-2023.

[35] Fam DWH, Palaniappan A, Tok AIY, Liedberg B, Moochhala SM. A review on technological aspects influencing commercialization of carbon nanotube sensors. Sens Actuators, B. 2011;157 (1):1-7.

[36] Lee S-H, Im JS, Kang SC, Bae T-S, In SJ, Jeong E, et al. An increase in gas sensitivity and recovery of an MWCNT-based gas sensor system in response to an electric field. Chem Phys Lett. 2010;497 (4):191-195.

[37] Cava CE, Salvatierra RV, Alves DCB, Ferlauto AS, Zarbin AJG, Roman LS. Self-assembled films of multi-wall carbon nanotubes used in gas sensors to increase the sensitivity limit for oxygen detection. Carbon. 2012;50 (5):1953-1958.

[38] Ndiaye AL, Varenne C, Bonnet P, Petit â, Spinelle L, Brunet Jrm, et al. Elaboration of single wall carbon nanotubes-based gas sensors: Evaluating the bundling effect on the sensor performance. Thin Solid Films. 2012;520 (13):4465-4469.

[39] Yao F, Duong DL, Lim SC, Yang SB, Hwang HR, Yu WJ, et al. Humidity-assisted selective reactivity between NO2 and SO2 gas on carbon nanotubes. J Mater Chem. 2011;21 (12):4502 - 4508.

[40] Chen G, Paronyan TM, Pigos EM, Harutyunyan AR. Enhanced gas sensing in pristine carbon nanotubes under continuous ultraviolet light illumination. Sci Rep. 2012;2:1-7.

[41] Battie Y, Ducloux O, Thobois P, Dorval N, Lauret JS, Attal-Trétout B, et al. Gas sensors based on thick films of semi-conducting single walled carbon nanotubes. Carbon. 2011;49 (11):3544-3552.

[42] Horrillo MC, Martí J, Matatagui D, Santos JP, Sayago I, Gutiérrez J, et al. Single-walled carbon nanotube microsensors for nerve agent simulant detection. Sens Actuators, B. 2011;157 (1):253-259.

[43] Wang F, Swager TM. Diverse Chemiresistors Based upon Covalently Modified Multiwalled Carbon Nanotubes. JACS. 2011 2011/07/27;133 (29):11181-11193.

[44] Shirsat MD, Sarkar T, Kakoullis J, Myung NV, Konnanath B, Spanias A, et al. Porphyrin-Functionalized Single-Walled Carbon Nanotube Chemiresistive Sensor Arrays for VOCs. The Journal of Physical Chemistry C. 2012 2012/02/09;116 (5): 3845-3850.

[45] Penza M, Rossi R, Alvisi M, Signore MA, Serra E, Paolesse R, et al. Metalloporphyrins-modified carbon nanotubes networked films-based chemical sensors for enhanced gas sensitivity. Sens Actuators, B. 2010;144 (2):387-394.

[46] Tisch U, Aluf Y, Ionescu R, Nakhleh M, Bassal R, Axelrod N, et al. Detection of Asymptomatic Nigrostriatal Dopaminergic Lesion in Rats by Exhaled Air Analysis Using Carbon Nanotube Sensors. ACS Chemical Neuroscience. 2011;3 (3):161-166.

[47] Ionescu R, Broza Y, Shaltieli H, Sadeh D, Zilberman Y, Feng X, et al. Detection of Multiple Sclerosis from Exhaled Breath Using Bilayers of Polycyclic Aromatic Hydrocarbons and Single-Wall Carbon Nanotubes. ACS Chemical Neuroscience. 2011 2011/12/21;2 (12):687-693.

[48] Wei L, Shi D, Ye P, Dai Z, Chen H, Chen C, et al. Hole doping and surface functionalization of single-walled carbon nanotube chemiresistive sensors for ultrasensitive and highly selective organophosphor vapor detection. Nanotechnology. 2011;doi: 10.1088/0957-4484/22/42/425501

[49] Wang Y, Yang Z, Hou Z, Xu D, Wei L, Kong ES-W, et al. Flexible gas sensors with assembled carbon nanotube thin films for DMMP vapor detection. Sens Actuators, B. 2010;150 (2):708-714.

[50] Slobodian P, Riha P, Lengalova A, Svoboda P, Saha P. Multi-wall carbon nanotube networks as potential resistive gas sensors for organic vapor detection. Carbon. 2011;49 (7):2499-2507.

[51] Cao CL, Hu CG, Fang L, Wang SX, Tian YS, Pan CY. Humidity Sensor Based on Multi-Walled Carbon Nanotube Thin Films. Journal of Nanomaterials. 2011;doi: 10.1155/2011/707303.

[52] Liu C-K, Wu J-M, Shih HC. Application of plasma modified multi-wall carbon nanotubes to ethanol vapor detection. Sens Actuators, B. 2010;150 (2):641-648.

[53] Im JS, Kang SC, Bai BC, Bae T-S, In SJ, Jeong E, et al. Thermal fluorination effects on carbon nanotubes for preparation of a high-performance gas sensor. Carbon. 2011;49 (7):2235-2244.

[54] Xie H, Sheng C, Chen X, Wang X, Li Z, Zhou J. Multi-wall carbon nanotube gas sensors modified with amino-group to detect low concentration of formaldehyde. Sens Actuators, B. 2011;doi: 10.1016/j.snb.2011.12.112.

[55] Koós AA, Nicholls RJ, Dillon F, Kertész K, Biró LP, Crossley A, et al. Tailoring gas sensing properties of multi-walled carbon nanotubes by in situ modification with Si, P, and N. Carbon. 2012;50 (8):2816-2823.

[56] Battie Y, Ducloux O, Patout L, Thobois P, Loiseau A. Selectivity enhancement using mesoporous silica thin films for single walled carbon nanotube based vapour sensors. Sens Actuators, B. 2012;163 (1):121-127.

[57] Talla JA. First principles modeling of boron-doped carbon nanotube sensors. Physica B: Condensed Matter. 2012;407 (6):966-970.

[58] Talla JA. Ab initio simulations of doped single-walled carbon nanotube sensors. Chem Phys. 2012;392 (1):71-77.

[59] Hamadanian M, Khoshnevisan B, Fotooh FK, Tavangar Z. Computational study of super cell Al-substituted single-walled carbon nanotubes as CO sensor. Computational Materials Science. 2012;58:45-50.

[60] Liu H, Ma H, Zhou W, Liu W, Jie Z, Li X. Synthesis and gas sensing characteristic based on metal oxide modification multi wall carbon nanotube composites. Appl Surf Sci. 2012;258 (6):1991-1994.

[61] Li W, Jung H, Hoa ND, Kim D, Hong S-K, Kim H. Nanocomposite of cobalt oxide nanocrystals and single-walled carbon nanotubes for a gas sensor application. Sens Actuators, B. 2010;150 (1):160-166.

[62] Tulliani J-M, Cavalieri A, Musso S, Sardella E, Geobaldo F. Room temperature ammonia sensors based on zinc oxide and functionalized graphite and multi-walled carbon nanotubes. Sens Actuators, B. 2011;152 (2):144-154.

[63] Jang DM, Jung H, Hoa ND, Kim D, Hong S-K, Kim H. Tin Oxide-Carbon Nanotube Composite for NOX Sensing. Journal of Nanoscience and Nanotechnology. 2012;12 (2):1425-1428.

[64] Leghrib R, Pavelko R, Felten A, Vasiliev A, Cané C, Gracia I, et al. Gas sensors based on multiwall carbon nanotubes decorated with tin oxide nanoclusters. Sens Actuators, B. 2010;145 (1):411-416.

[65] Wongchoosuk C, Wisitsoraat A, Phokharatkul D, Tuantranont A, Kerdcharoen T. Multi-Walled Carbon Nanotube-Doped Tungsten Oxide Thin Films for Hydrogen Gas Sensing. Sensors. 2010;10 (8):7705-7715.

[66] Van Hieu N, Duc NAP, Trung T, Tuan MA, Chien ND. Gas-sensing properties of tin oxide doped with metal oxides and carbon nanotubes: A competitive sensor for ethanol and liquid petroleum gas. Sens Actuators, B. 2010;144 (2):450-456.

[67] Leghrib R, Felten A, Pireaux JJ, Llobet E. Gas sensors based on doped-CNT/SnO2 composites for NO2 detection at room temperature. Thin Solid Films. 2011;520 (3): 966-970.

[68] Yun J, Im JS, Kim H-I, Lee Y-S. Effect of oxyfluorination on gas sensing behavior of polyaniline-coated multi-walled carbon nanotubes. Appl Surf Sci. 2012;258 (8): 3462-3468.

[69] Mangu R, Rajaputra S, Singh VP. MWCNT–polymer composites as highly sensitive and selective room temperature gas sensors. Nanotechnology. 2011;doi: 10.1088/0957-4484/22/21/215502.

[70] Sayago I, Fernández MJ, Fontecha JL, Horrillo MC, Vera C, Obieta I, et al. New sensitive layers for surface acoustic wave gas sensors based on polymer and carbon nanotube composites. Procedia Engineering. 2011;25 (0):256-259.

[71] Sayago I, Fernández MJ, Fontecha JL, Horrillo MC, Vera C, Obieta I, et al. New sensitive layers for surface acoustic wave gas sensors based on polymer and carbon nanotube composites. Sens Actuators, B. 2012;doi:10.1016/j.snb.2011.12.031.

[72] Sayago I, Fernández MJ, Fontecha JL, Horrillo MC, Vera C, Obieta I, et al. Surface acoustic wave gas sensors based on polyisobutylene and carbon nanotube composites. Sens Actuators, B. 2011;156 (1):1-5.

[73] Viespe C, Grigoriu C. Surface acoustic wave sensors with carbon nanotubes and SiO2/Si nanoparticles based nanocomposites for VOC detection. Sens Actuators, B. 2010;147 (1):43-47.

[74] Yun S, Kim J. Multi-walled carbon nanotubes-cellulose paper for a chemical vapor sensor. Sens Actuators, B. 2010;150 (1):308-313.

[75] Kumar B, Castro M, Feller JF. Poly (lactic acid)-multi-wall carbon nanotube conductive biopolymer nanocomposite vapour sensors. Sens Actuators, B. 2012;161 (1): 621-628.

[76] Yuana CL, Chang CP, Song Y. Hazardous industrial gases identified using a novel polymer/MWNT composite resistance sensor array. Mater Sci Eng, B. 2011;176 (11): 821-829.

[77] Li W, Hoa ND, Kim D. High-performance carbon nanotube hydrogen sensor. Sens Actuators, B. 2010;149 (1):184-188.

[78] Lvova L, Mastroianni M, Pomarico G, Santonico M, Pennazza G, Di Natale C, et al. Carbon nanotubes modified with porphyrin units for gaseous phase chemical sensing. Sens Actuators, B. 2011:doi:10.1016/j.snb.2011.1005.1031.

[79] Lu Y, Meyyappan M, Li J. Trace Detection of Hydrogen Peroxide Vapor Using a Carbon-Nanotube-Based Chemical Sensor. Small. 2011;7 (12):1714-1718.

[80] Lu Y, Meyyappan M, Li J. A carbon-nanotube-based sensor array for formaldehyde detection. Nanotechnology. 2011;doi:10.1088/0957-4484/22/5/055502.

[81] Ruiz A, Arbiol J, Cirera A, Cornet A, Morante JR. Surface activation by Pt-nanoclusters on titania for gas sensing applications. Materials Science and Engineering: C. 2002;19 (1-2):105-109.

[82] Ghasempour R, Mortazavi SZ, Iraji zad A, Rahimi F. Hydrogen sensing properties of multi-walled carbon nanotube films sputtered by Pd. Int J Hydrogen Energy. 2010;35 (9):4445-4449.

[83] Randeniya LK, Martin PJ, Bendavid A. Detection of hydrogen using multi-walled carbon-nanotube yarns coated with nanocrystalline Pd and Pd/Pt layered structures. Carbon. 2012;50 (5):1786-1792.

[84] Su P-G, Chuang Y-S. Flexible H2 sensors fabricated by layer-by-layer self-assembly thin film of multi-walled carbon nanotubes and modified in situ with Pd nanoparticles. Sens Actuators, B. 2010;145(1):521-526.

[85] Zilli D, Bonelli PR, Cukierman AL. Room temperature hydrogen gas sensor nanocomposite based on Pd-decorated multi-walled carbon nanotubes thin films. Sens Actuators, B. 2011;157 (1):169-176.

[86] Ju S, Lee JM, Jung Y, Lee E, Lee W, Kim S-J. Highly sensitive hydrogen gas sensors using single-walled carbon nanotubes grafted with Pd nanoparticles. Sens Actuators, B. 2010;146 (1):122-128.

[87] Rumiche F, Wang HH, Indacochea JE. Development of a fast-response/high-sensitivity double wall carbon nanotube nanostructured hydrogen sensor. Sens Actuators, B. 2012;163 (1):97-106.

[88] Sadek AZ, Bansal V, McCulloch DG, Spizzirri PG, Latham K, Lau DWM, et al. Facile, size-controlled deposition of highly dispersed gold nanoparticles on nitrogen carbon nanotubes for hydrogen sensing. Sens Actuators, B. 2011;160 (1):1034-1042.

[89] Penza M, Rossi R, Alvisi M, Serra E. Metal-modified and vertically aligned carbon nanotube sensors array for landfill gas monitoring applications. Nanotechnology. 2010;doi:10.1088/0957-4484/21/10/105501.

[90] Penza M, Rossi R, Alvisi M, Suriano D, Serra E. Pt-modified carbon nanotube networked layers for enhanced gas microsensors. Thin Solid Films. 2011;520 (3):959-965.

[91] Zhou X, Tian WQ, Wang X-L. Adsorption sensitivity of Pd-doped SWCNTs to small gas molecules. Sens Actuators, B. 2010;151 (1):56-64.

[92] Li K, Wang W, Cao D. Metal (Pd, Pt)-decorated carbon nanotubes for CO and NO sensing. Sens Actuators, B. 2011;In Press, Corrected Proof:doi: 10.1016/j.snb. 2011.1006.1068.

Chapter 4

Carbon Nanotube Composites for Electronic Interconnect Applications

Tamjid Chowdhury and James F. Rohan

Additional information is available at the end of the chapter

1. Introduction

1.1. Electronic interconnect

In the electronics industry interconnect is defined as a conductive connection between two or more circuit elements. The interconnect connects elements (transistor, resistors, etc.) on an integrated circuit or between components on a printed circuit board. The main function of the interconnect is to contact the junctions and gates between device cells and input/output (I/O) signal pads. These functions require specific material properties. For performance or speed, the metallization structure should have low resistance and capacitance. For reliability, it is important to have the capability of carrying high current density, stability against thermal annealing, resistance against corrosion and good mechanical properties.

Over the past 40 years the continuous improvements in microcircuit density and performance predicted by Moore's Law has led to reduced interconnect dimensions. According to Moore's law the number of transistors incorporated in a chip will approximately double every 18-24 months. The interconnect length increases with each generation, leading to higher resistances, while the distance between the adjacent interconnects decreases, leading to increase capacitance. Previously Al interconnect was used for VLSI processing [1]. Al and its alloys, suffer from the problems of high resistance-capacitance (RC) delay (a "time-delay" between the input and output, when a signal or voltage is applied to a circuit), poor electromigration resistance and poor mechanical properties for application in ultra-large-scale integrated (ULSI) circuits [2].

Table 1 shows the comparison of different metals resistivity at room temperature. It can be seen from the table that only three metals have lower resistivity than Al, namely Ag, Au and Cu. Ag has the lowest resistivity but it has poor electromigration reliability. Electromigration is the transport of material caused by the gradual movement of the ions in a conductor due to the momentum transfer between conducting electrons and diffusing metal atoms. The resistivity of Cu is 1.67 μΩ.cm, which is about 40% better than Al. The self-diffusivity (the spontaneous movement of an atom to a new site in a crystal of its own species) of Cu is also the smallest among the four metals, resulting in improved reliability [3, 4].

Metal	Bulk resistivity μΩ.cm
Ag	1.63
Cu	1.67
Au	2.35
Al	2.67
W	5.65

Table 1. Comparison of the bulk resistivity for different metals.

Table 2 shows a comparison of the activation energy (the minimum energy required for movement of an atom from a lattice position in a crystal) and melting temperature of Al vs. Cu. It can be seen from this table that Cu is a more reliable metal than Al with more energy required for diffusion of Cu atoms. The reason that Cu had not been used much earlier than its introduction in 1997 was because of device reliability concerns and processing difficulties. Cu diffuses rapidly through SiO_2 in the presence of an electric field [5]. This causes degradation of transistor reliability by increasing metallic impurity levels in the Si. Another problem with Cu is that it oxidises at low temperatures but without self-passivation [6]. Cu is also difficult to etch unlike Al. This means that the classical approach where metal is deposited over the entire surface, structures created in the metal and finally infilled with dielectric (oxide) cannot be followed with Cu.

Interconnect Metal	Melting Point °C	Ea for lattice diffusion eV	Ea for grain boundary diffusion eV
Al	660	1.4	0.4 – 0.8
Cu	1083	2.2	0.7 – 1.2

Table 2. Comparison of the active energy for diffusion of Al vs. Cu.

To overcome the problems of Cu integration the inter-level dielectric (ILD) is first deposited and patterned to define "trenches" into which the metal lines of the interconnect will be placed. A thin layer of barrier material (typically refractory metals or their alloys) is deposited generally using a physical vapour deposition (PVD) process. This layer covers the entire surface to act as a barrier to Cu diffusion. After the deposition of the barrier layer the Cu interconnect is deposited. This can be achieved by conventional methods such as physical vapour deposition (PVD) [7] and chemical vapour deposition (CVD) [8]. However, PVD presents poor step coverage in sub-micrometer dimension vias and trenches. This technique deposits blanket films which would require further patterning. CVD also requires the use of combustible and toxic precursors at elevated temperatures which has limited the development of Cu deposition by CVD. In 1997 IBM developed the electrodeposition technique (dual damascene) for Cu metallization [9]. Electrodeposition has become the standard method for Cu metallization with demonstrated uniformity, gap filling ability and low processing temperatures. In the dual damascene technique, lines and vias can be filled with electrodeposited Cu at the same time. Fig. 1 shows a schematic diagram of via filling with Cu and the requirement to achieve superfilling or bottom up deposition through the use of suitable additives in the plating bath rather than subconformal or conformal which result in voids or seams in the Cu. [10-12].

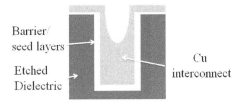

Figure 1. Cross section schematic of interconnect trench or via showing 'super-filling' or 'bottom-up filling' of features through the use of specific plating bath additives for optimum void-free profile evolution in damascene processing [9].

Semiconductor manufacturers have adopted the electroplating technique for Cu interconnect deposition in electronic devices and continue to work on miniaturization of device and feature sizes. Fig. 2 shows cross-sections from the International Technology Roadmap for Semiconductors (ITRS) of a typical microprocessor and application specific integrated circuit where the interconnect of different lines and vias between two adjacent layers are filled with Cu. As the feature sizes decrease and consequently the operating currents increase, electromigration becomes a serious issue once more [13].

The effect is important in applications where high direct current densities are used, such as in high performance processors. Grain boundaries are the fastest diffusion path for Al electromigration (activation energy 0.6 eV for grain boundary diffusion and 1 eV for interface diffusion) but an interface is the fastest diffusion path for Cu (activation energy 1.2 eV for grain boundary diffusion and 0.7 eV for interface diffusion) [14, 15]. The difference in electromigration mechanism drives different focus areas for Cu and Al reliability improvement. The damascene process requires the removal of overdeposited Cu by chemical mechanical polishing (CMP). The CMP produced top Cu surface is the fast Cu diffusion path which needs to be reliably capped. A nonconductive barrier layer is generally applied as the cap layer (silicon nitride, silicon carbide, nitride silicon carbide etc) is used to cover the top surface of the Cu line. However, there are some issues with using dielectric caps to passivate

Cu. As devices become smaller, the current density through the interconnect increases leading to the requirement for better electromigration resistance. The dielectric cap generally has a higher dielectric constant than the interlevel dielectric, resulting in an increase in line-to-line capacitance. Improved Cu electromigration resistance was reported when Cu lines were protected with thin conductive surface capping layers of self-aligned electrolessly deposited CoWP or CoSnP etc [16, 17]. Diffusion barrier layers such as Ta or TaN for Cu metallization act as redundant layers for current shunting as well as for uniform Cu seed deposition. It was reported that Cu vias are the weak link in the interconnect metallization [14]. The Cu via connects directly to the Cu metal below. If a void forms in the Cu underneath the via, there is no redundant layer available for current shunting. This is the primary cause of early failure distribution in Cu interconnects. For the 22 nm technology node or below, the interconnect metal should have current carrying capability of more than 10^7 A/cm^2 to overcome the electromigration issue but Cu is limited to 10^7 A/cm^2.

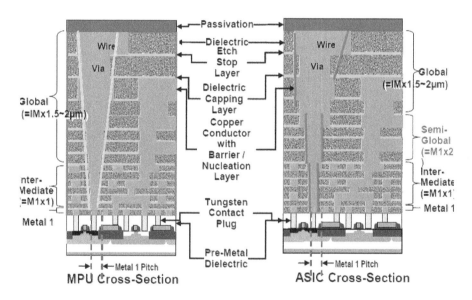

Figure 2. Typical cross section illustrating hierarchical scaling methodology [ITRS technology road map, 2011 update].

1.2. Carbon nanotubes

Carbon nanotubes (CNTs) have unique electrical, thermal and mechanical properties [18]. They can carry an electrical current density of $\sim 4 \times 10^9$ A cm^{-2}, which is three orders of magnitude higher than Cu [19]. CNT's have high aspect ratio and the mean free path of the carriers (or the probability of an electron transmitted from at one end of the CNT to the other without phonon scattering or other thermal effects) in the CNT at 10 µm is much longer than any metal (e.g. 40 nm for Cu). In addition the covalent C-C bonding between the neigh-bouring atoms in the CNT is one of the strongest bonds reported in the literature [20] and C atoms will not migrate even under very high current density (activation energy 7.7 eV for atom movement). Thus the electromigration resistance of CNT is much better than other in-terconnects material such as Al and Cu. Because of these advantages over Cu, CNTs require consideration as the next generation interconnect material for specific applications, such as through silicon vias (TSV) for stacked die.

However there are still significant scientific and engineering challenges to incorporate CNTs in devices. CNTs can deposit inside of vias on suitable catalyst like zeolite [21]. It is necessa-ry to ensure the selective growth of metallic CNTs in vias and lines to achieve better electri-cal conductivity. Alternatively CNTs can be first synthesized in a powder form and metallic CNTs separated from bulk growth CNTs (mixture of metallic and semiconducting). After that metallic CNTs would need to be transferred onto specific wafer locations. The scale of this task is obvious when considering that there are billions of transistors in a microproces-sor and the placement of CNTs inside of all vias and trenches on the wafer is unlikely.

	Single CNT	Cu at 22 nm node
Maximum Current density (A/cm^2)	1×10^9	1×10^7
Electrical conductivity (S/m)	10^6-10^7	6×10^7
Thermal coefficient of resistivity (/°C)	-1.5×10^{-3}	4×10^{-3}
Thermal conductivity (W/m K)	6,000	400
Coefficient of thermal expansion (ppm/°C)	-1.5	17
Activation energy (eV)	7	2

Table 3. Comparison of the properties of single walled CNTs vs. Cu.

The contact resistance between CNTs and metal is large (≥ 1 kΩ). The minimum resistance for a ballistic single-walled CNT is ~ 6.5 kΩ, Therefore, relatively dense arrays of nanotubes will be needed to replace Cu interconnects and these arrays will still only show reduced re-sistance by comparison with Cu interconnect for line lengths ≥ 1 μm. The ITRS [22] therefore predicts for the 22 nm node an estimated resistance for a 17 nm x 38.5 nm x 1.5 μm Cu inter-connect is R_{Cu} 145 Ω. An ensemble of ~ 45, 1 nm diameter defect-free metallic CNTs with mean inter-tube separations ~ 4 nm in a trench of these dimensions would have the same total resistance as the Cu line.

Intrinsic voids between CNTs significantly reduce the electrical and thermal conductivity and bring reliability challenges for the use of CNTs as interconnects. Contact resistance between CNTs and interconnect metal like Cu becomes the dominant source of electrical and thermal resistance which significantly reduces the benefits of CNTs. The potential use of CNTs alone as interconnect in semiconductor manufacturing is still open to debate at this time.

1.3. Metal CNT composites

To overcome some of the interconnect issues described above metal-CNT composites can be an alternative candidate material for future interconnects. The composite material would increase the contact area between vias and interconnect lines. There is also less chance of intrinsic voids between CNTs as they would be metal filled. Among the different metals Cu is the best choice at this moment to use as a composite material with CNTs for interconnects applications because of superior electrical and thermal conductivity. Chai et al [23,24] and Yoo et al [25] reported that Cu fills the voids between neighbouring CNTs which results in a more densely packed structure. They reported that the addition of Cu increases the contact area between the nanotube (1 D) and the substrate (3 D contact) making it a mechanically strong material that can sustain high electrical or thermal stress cycling. To obtain superior properties of metal-CNT composites, it is necessary to achieve a homogeneous dispersion of CNT throughout the metal matrix. It is also necessary that the composites should be void free to obtain better electrical and thermal conductivity.

Cu/CNT composites can be prepared by powder metallurgy, electroless plating or electrodeposition techniques [25, 26]. Among these methods electrochemical routes are relatively straightfoward methods to produce defect free nanocomposites [24, 25]. Powder metallurgy requires sintering at elevated temperatures that may damage CNT's and the difficulty of composite placement remains an issue. Chen et al [26] observed a clear separation of CNTs and Cu matrix composites deposited by powder metallurgy. To achieve optimum performance, CNTs need to be well-dispersed and aligned parallel rather than randomly oriented in the Cu matrix. Hjortstam et al [27] estimated the increase of effective conductivity as a function of the volume fraction of CNT in a Cu matrix. Their calculation showed that 30-40% CNT is needed in the composite with a resistivity 50% lower than for Cu. Liu et al [28] found that electrical sheet resistance is lower in Cu/CNT composite films than Cu and also decreases due to annealing at 200 - 300°C.

Improved electromigration resistance is expected to result from the location of the alloy element at grain boundaries to prevent movement of Cu at those vulnerable points, which may lead to wiring voids (opens) or hillocks (shorts) during operation [25]. Cu/CNT composites may also improve thermal conductivity of lines and vias which also increases electromigration resistance. Chai et al [24] reported that the Cu/CNT composite vias have lower electrical resistance than that of vias with CNT only. Their electromigration test results showed that the void growth rate for a Cu/CNT composite strip was four times lower than that of pure Cu strip. Their electromigration test of Cu and Cu/CNT composites which were carried out in the temperature range of 100 to 250°C and current density from 5×10^5 to 2×10^6 A/cm^2 using a conventional Blech-Kinsborn test structure showed that longer strips had larger void length, while no void formation was detected in the strips below 40 μm. Below the critical length the electromigration flux is balanced by the opposing backflow generated by the stress gradient in the test strip.

Yoo et al [25] fabricated Cu/MWCNT composite films by a pulsed electrodeposition technique with additives and obtained a dense structure without any voids. Their microstructure analysis showed that most of the MWCNTs exist at the Cu grain boundaries and cross-linked each other. They reported that C content in the composite increased by increasing CNT concentration in the bath but it decreased with annealing. Chai [29] et al reported that the mechanical strength of Cu/CNT nanocomposite was three times higher than that of pure Cu. Chen et al [26] observed good interfacial bonding between CNT and Cu when the nanocomposites were codeposited by electrodeposition. They reported that for Cu/SWCNT nanocomposites, the radial breathing mode (RBM) in the Raman was absent and the tangential or G-band had shifted and widened. Recently several patents on the metal/CNT composites codeposited by electrodeposition have been filed [30-32]. The comparison of electrical resistivity of Cu and Cu/SWCNT film which was reported by Chan [30] are shown in table 4.

	electrodeposited Cu (thickness = 10.5 μm)	Cu/CNT composite (thickness = 22 μm)
Resistivity (μΩ.cm)	1.72	1.22
Sheet resistance (mΩ/sq)	1.64	0.56

Table 4. Comparison of the electrical resistivity of electrodeposited Cu/SWCNT composite film vs. Cu alone [30].

1.4. Chlorosulphonic acid for CNT dissolution

Recently Davis et al [33] reported that CNTs can dissolve spontaneously in chlorosulphonic acid solution up to 0.5 wt % [5 g/l], which is much higher than previously reported in other acids (up to 80 mg/l). They reported that at higher concentrations, they form liquid-crystal phases that can be processed into fibres and sheets of controlled morphology. Their proposed phase diagram helps to identify the optimal starting fluid composition and determine micro and macrostructure of fibres and films such as plated fibres, straight fibres and smooth films. Plated fibres have potential application for hydrogen storage and sensors because of high surface area. Straight fibres are of interest for structural reinforcement and smooth, dense films for electrical applications such as electrically conductive thin films.

1.5. Purification and functionalization of carbon nanotubes

A significant problem in dealing with CNTs is the difficulty to separate them as the individual CNTs form bundles due to van der Waals attractive forces. Also in all of the synthesis techniques several impurities like catalyst particles, amorphous carbon etc. are also present in the bundles of CNTs. These impurities may deteriorate the properties of CNTs. To prepare stable and homogeneous dispersions of CNTs considerable efforts have been made [34-39] but the solubility of CNTs in water or organic solvent is relatively low. At room temperature the solubility of CNTs is in the range of 60 to 80 mg/l [34]. In order to achieve better stabilization, CNTs require additional hydrophilic groups directly on the CNT walls or provided by surfactant molecules to impart ionic charge on the CNTs [40-42]. The most common hydrophilic groups are $-OH-$, $-COOH-$, $-SO_3-$, $-NH_2-$. Functionalization of CNTs can be an important factor to manipulate the properties of CNTs. With functionalization CNTs mau be more easily separated. Several methods have been suggested for the purification and functionalization of CNTs mainly based on covalent and noncovalent functionalization. Functionalized CNTs are easily dispersed and highly ionized in contract with water [43].

1.5.1. Covalent functionalization of carbon nanotubes

Several methods have been suggested for covalent functionalization of CNTs. The most common technique is to functionalize CNTs in concentrated acid by refluxing. In this process raw materials are sonicated followed by refluxing at 120-130°C. This process requires long processing times. After cooling at room temperature, the mixture is then centrifuged, leaving a black precipitate and a clear brownish yellow supernatant acid. Ko et al [44] reported that the presence of metal impurities in the MWCNTs is reduced significantly using this method. The purification process usually requires two repeat processing steps. The first step is acid reflux which washes metal catalyst and carbon impurities and the second step is annealing which burns the defective tubes and carbon particles. Ko et al [44] also used a microwave oven technique to purify MWCNTs. Chen and Mitra [45] reported that MWCNTs were less reactive and had lower solubility than the SWNTs. Li and Grennberg [46] also found that microwave heating is highly useful for side wall functionalization of MWCNTs.

Lau et al [47] reported that the electrical conductivity of MWCNTs increased with different functionalization techniques such as oxidation, acid reflux, dry UV-ozonolysis. They explained that the new functionalized groups increase the number of bands near the Fermi level, promoting electron transfer between the carbon atoms. They have claimed that CNT functionalization by UV-ozonolyzed technique significantly increases the electrical conductivity of CNTs. Agarwal et al [48] reported that controlled defect creation could be an attractive strategy to induce an electrical conductivity increase in MWCNTs. They reported that the outermost shell of MWCNTs is semiconducting so it is difficult to make electrical contacts to the inner shells of MWCNTs. Functionalization of CNTs may promote cross-shell bridging via sp^3 bond formation. They proposed that intershell bridging facilitates charge carrier hopping to inner shells which can serve as additional charge carrier transport pathways. Tantang et al [49] also reported that acid treatment increases the conductivity of CNT electrodes.

1.5.2. Non-covalent functionalization of carbon nanotubes

Covalent functionalization may deteriorate the unique ionic properties of CNTs by the formation of new covalent bonds on the CNTs wall. To overcome this disadvantage non-covalent functionalization mainly based on polymer surfactant interaction was developed that can disperse nanotubes easily but not degrade the CNT's unique properties [50]. The proposed mechanism for this solubilisaiton is through an individual CNT being wrapped by the polymer which acts as a surfactant in the solution to achieve separation. A surfactant is a wetting agent which lowers surface tension of liquids. It is usually an organic compound that contains both hydrophobic and hydrophilic groups. As a result, they are soluble both in organic solvents and water. Surfactants are classified based on the presence of a charged group. The head of an ionic surfactant carries a net charge. If the charge is negative, the surfactant is more specifically called anionic; if the charge is positive, it is cationic. Ionic surfactants not only separate individual CNTs but also carry charge to the surface of CNTs so that the CNTs can be codeposited with Cu by electrodeposition. Examples of surfactants which have been investigated are given below.

1. Nafion® as a surfactant

In our study we primarily used nafion, a polymer surfactant for the dispersion of CNT bundles. It is a sulfonated tetrafluorethylene co-polymer with ionic properties which bears a polar side chain ($-SO_3H$) and hydrophobic backbone ($-CF_2-CF_2$). It has unique ionic properties because of the incorporation of perfluorovinyl ether groups terminated with sulfonate groups onto a tetrafluoroethylene (Teflon) backbone. The hydrophobic backbone strongly anchors to the hydrophobic side-wall of CNTs. On the other hand the polar side-chain of the polymer imparts sufficient ionic charge to the CNT surface which enhances the solubility of CNTs in liquid solvents. Fig. 3 shows the chemical structure of nafion.

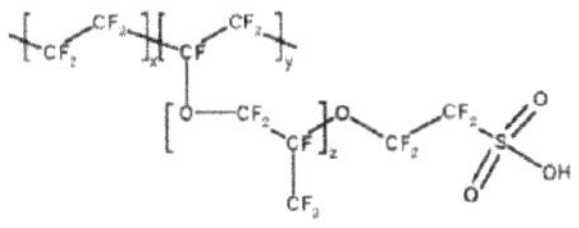

Figure 3. Chemical structure of Nafion.

2. CTAB as a surfactant

Cetyl trimethyl ammonium bromide (CTAB) is a cationic surfactant. Chen et al [51, 52] used CTAB for the dispersion of CNTs to prepare CNTs/Ni composites by an electroless deposition technique. The chemical structure of CTAB is shown in fig. 4 (a). As a cationic surfactant it can make the CNT surface positively charged to assist the codeposition of CNT on the cathodic surface [52]. The CNT with negative charge readily adsorbs the cationic surfactant. This adsorption develops a net positive charge on the CNT, which prevents them from agglomerating and leads to electrostatic attraction to the cathode surface with negative potential [51]. The net positive charge on the CNT increases the amount of CNT in the deposits. To calculate the volume fraction of CNTs, they dissolved the deposits in nitric acid. The CNTs in the deposits were filtered and the quantity of the CNTs in the deposits determined [52]. They reported that the content of CNTs in the deposit increases with an increase of CNT concentration in the bath, up to a maximum value at the CNT concentration of 1.1 g/l and then decreases. They explained this as a result of the CNT agglomeration in solution at higher concentration which reduces the content of CNT in the deposit. They also reported that the saturation concentration increases with decrease of length of CNTs because the longer CNTs tend to agglomerate more readily.

3. SDS as a surfactant

Sodium dodecyl sulfate (SDS) is an anionic surfactant used to improve the surface uniformity of the composite deposit [53]. The chemical structure of SDS is shown in fig. 4 (b).

Figure 4. Chemical structure of (a) CTAB and (b) SDS.

A comparison of hardness and XRD patterns of Ni/CNT composites by using SDS and CTAB surfactant in the bath with Ni was performed [53]. The hardness changes for the composite films and depends on the concentration of CNTs in the bath as well as surfactant. The composite from the CTAB bath showed an increase of hardness unlike that of the composite from the SDS containing bath. It can be also seen from the XRD data that (111) is the preferred plane of Ni when the bath contains CTAB like pure Ni deposition. On the other hand, SDS in the bath reduces the preferred Ni (111) orientation significantly in the composite. A summary of surfactants which are commonly used for the dispersion of CNTs are reported in table 5.

CNT	CNT: g/l	Surfactant	Surfactant: g/l	Composites	References
MW	0.3	CTAB	0.6	Ni/CNT	[53]
MW	0.1	$Mg(NO_3)_2$		Cu/CNT	[26]
MW	6	PA	0.5	Cu/CNT	[23]
SW	3	CTAC	3	Cu/CNT	[54]
MW	1	CTAB	4	Cu/Zn	[55]
SW	0.002	Nafion	0.4	Nafion/CNT	[56]
MW	0.6	Gelatine	0.4	CNT/Cu	[57]
MW	2	CTAB		Ni/CNT	[58]
SW	2			Cu/CNT	[24]
MW	1	Nafion	0.01	Nafion/CNT	[59]
MW	6	PA	0.1	Cu/CNT	[60]
MW	1			Ni-Co/CNT	[61]
MW	1	Nafion	5	Nafion/CNT film	[62]
SW	0.2	Nafion		Nafion/CNT	[63]
SW	0.05	SDS	0.1	PVP/CNT	[50]

Table 5. Literature review of CNT dispersion surfactants.

1.6. Analysis of carbon nanotubes in deposit

There is very little in the literature on quantifying CNT content in the deposits [48, 51, 52]. Arai et al [60, 64, 65] measured the content of multi-walled CNTs in the electrodeposited composite film by dissolving the deposit in hot nitric acid. The CNTs in the nitric acid solution were filtered, dried and weighed. Osaka et al [66] reported the carbon content in the deposit was analyzed by the combustion infrared absorption method (CS 444, LECO) as this element analyzer is capable of analysing for carbon. The summary of CNT content in the deposit obtained from the literature is shown in table 2.5.

Bath	Surfactant	CNT g/l	Weight % CNT in deposit	Reference
Ni	PA	6	Up to 1	[64]
Cu	PA	2	0.4	[60]
Ni	SDS	0.3	Up to 7	[53]
Cu	PA	Up to 6	Up to 2.5	[25]

Table 6. Literature data of CNT content in the plating bath and deposit.

2. Cu and Cu/CNT pillars for flip chip interconnect assembly

Historically, IC chips have been electrically connected to the substrate by a wire bond method. In this method, the chip faces up and is attached to the package via wires. This connection has limited electrical performance and reliability problems in addition to requiring pad location at the edge of the die. Flip chip, also known as 'Controlled Collapse Chip Connection, C4', replaced the traditional wire bond method. In this method, solder bumps are deposited on the chip pads over the full area of the top side of the wafer during the final wafer processing step. In order to mount the chip to external circuitry (e.g., a circuit board or another chip or wafer), it is flipped over so that its top side faces down, aligned to the substrate and then the solder is reflowed to complete the interconnect. Generally, Sn-Pb based solder bumps have been used in flip chip packaging to connect chips to external circuitry. According to the International Technology Roadmap for Semiconductors the total number of I/Os will reach up to 10,000 cm² chip area by 2014 which require finer interconnect with a pitch size less than 20 μm. To fabricate such fine pitch interconnect, conventional solder bump requires fine solder deposition or paste particle which are not readily available [67]. It is also important to reduce lead-based solders for environmental concerns (RoHS compliance). As circuit density increases, devices are also more vulnerable to non-uniform thermal distribution.

Cu has higher thermal conductivity than most binary or ternary solders. Cu bumps in flip chip assembly offer increased reliability, extended temperature range capability, greater mechanical strength, higher connection density, improved manufacturability, better electrical and heat dissipating performance over Pb-Sn solder. It is also less expensive and decreases the amount of solder needed to create bumps. Cu pillars do not change shape during reflow so they do not encounter any volumetric redistribution which can lead to voids in the solder. Because of these advantages the semiconductor industry is adopting Cu pillar bump by electrodeposition for flip-chip attachment to replace the typical Pb solder [67, 68]. Power and thermal non-uniformity in devices are increasing steadily with each new device generation leading to serious concerns for the industry regarding thermal issues. Mechanical stress on Cu bumps generated by the difference in thermal expansion coefficients between the chip and the substrate materials can lead to device failures. This differential thermal expansion also creates shearing forces at the bump. As a result bumps are most vulnerable to damage. Repeated thermal expansion and contraction leads to fatigue cracking of the bump.

Cu/CNT composites could be a suitable candidate material to resolve these issues for next generation flip chip assembly. CNTs have high mechanical strength (10-60 GPa, c.f. Cu 70 MPa) and thermal conductivity (>3000 W/m.K, c.f. Cu 400 W/m.K) which may alleviate the issues related to die degradation and non-uniform temperature distribution in the pillars. CNTs have a negative temperature coefficient of resistivity (- 1.5 x10^{-3}/$^{\circ}$C, c.f. Cu + 4 x10^{-3}/$^{\circ}$C) and low coefficient of thermal expansion (- 1.5 ppm/$^{\circ}$C, c.f. Cu + 17 ppm/$^{\circ}$C) which can make the Cu/CNT composites material more reliable against thermal cycling and fatigue with less risk of stress induced failure. Typical photolithography techniques can be utilised to fabricate Cu/CNT pillar bumps on chip. Arai et al [64] recently demonstrated Cu/CNT pillar emitters deposited by electrodeposition on a patterned substrate.

3. Cu and Cu/CNT in through Si via (TSV) for 3D interconnect

Cu electrodeposition in TSV features is a key component of new 3D integration approaches that are of great interest in the semiconductor industry [69]. 3D integration increases performance and lowers power consumption due to reduced length of electrical connections. Cu has been selected as the TSV interconnect because of its low electrical resistance and compatibility with conventional multilayer interconnection in large-scale integration (LSI) and back-end processes. The key challenges for TSV plating processes are to fill the vias across the entire wafer and to complete the fill as fast as possible to minimize cost. TSV interconnect shortens the interconnect requirements and reduces signal delay. However, it is difficult to fill high aspect ratio vias without voids through conventional damascene electroplating. Perfect filling without voids is required to minimise interconnect failure and reliability issues. TSVs have been extensively studied because of their ability to achieve chip stacking for enhanced system performance. This is a very promising technology that may replace wire bonding in chips or single chip solder bumping. Metal filled TSVs allow devices to be connected using a 3D approach [69]. Cu is the best low cost conductor for TSV interconnect and an extension of the damascene plating in smaller features. Recently enormous attention has been given to bottom up filling of TSVs to fill high aspect ratio vias without voids like conventional damascene electroplating [70-72]. However, there are key issues that need to be resolved, such as process reliability, electrical continuity and thermal management. TSVs should have the ability to maintain operation over a wide range of temperatures and to withstand these temperatures in a cyclic manner. The TSV material properties must include mechanical strength, good thermal conductivity and stability with thermal cycling.

The key issues for 3D integration are process reliability, die degradation, electrical continuity, bump to pad electrical contact and thermal management. Temperature cycling and thermal shock accelerate fatigue failures. Also, non-uniform temperature distribution may influence the operation of circuits and sensing elements dramatically. Stress fields resulting from differential thermal expansion of Cu-based TSV may cause serious problems. The reliability problems of high aspect ratio TSV interconnect may be alleviated by the codeposition of carbon nanotubes (CNTs) with Cu as a suitable composite material. CNTs have high mechanical strength, thermal conductivity and low coefficient of thermal expansion which may alleviate the issues related to die degradation and non-uniform temperature distribution in the TSVs.

4. Experimental methods

In this work the Cu/CNT composites codeposition process was assessed and the deposited materials characterised. Electrochemical analysis of the deposition requires an analysis of the nucleation and growth characteristics for the candidate materials. MWCNTs have been added to the typical Cu sulphate plating bath to achieve homogeneous Cu/MWCNT composites. Here, we will report electrochemical analysis and kinetics of electrodeposited Cu when MWCNTs were present in the bath. Solubilisation or suspension of the CNTs in the Cu bath is also a key requirement. Composite plating bath chemistries for Cu/CNT deposition were investigated. The influence of typical additives in the Cu bath on the deposit characteristics was determined for optimised electrodeposition in vias and trenches. The influence of different surfactants on the deposition and electrical properties of composite films was also analyzed. Cu and Cu/CNT composites were electrodeposited on planar and structured substrates. Microstructure characterization of the deposit employed scanning electron microscopy (SEM), focussed ion beam microscopy (FIB) and x-ray diffraction (XRD). The sheet resistance of Cu/CNTs film and changes due to self-annealing and high temperature annealing were monitored by 4 point probe resistivity techniques. Cu/CNT composites were also deposited in test structures. After chemical mechanical polishing of the test structures, the line resistance was measured using a Cascade probe station.

The amount of CNTs in the deposit was determined by dissolving the deposit in a concentrate HNO_3 solution. The Cu/CNT films were deposited on 1 cm X 1 cm thin film sputtered Cu on Si. The deposition current was 1 A and deposition time was 1 h. The concentration of CNTs in the bath was 10 or 100 mg/l. After deposition, the sample was dipped in hot concentrate acidic solution (65% HNO_3, 65°C). The diluted acid solution was then vacuum filtered using PTFE filter paper. The filtration process was repeated at least 5 times to ensure all CNTs were left on filter residue. After filtration, the PTFE membrane was dried in an oven at 80°C for at least 30 minutes to ensure the membrane was completely dried. The weight difference of the PTFE membrane before and after filtration gives the amount of

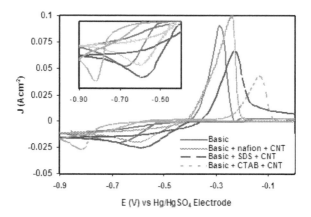

Figure 5. CVs of Cu and Cu/CNTs deposition on a glassy carbon electrode (scan rate: 0.1 V/s) from 0.24 mol dm⁻³ CuSO₄ + 1.8 mol dm⁻³ H₂SO₄ with/without CNTs and different surfactants in the bath.

CNTs in the deposit. The amount of CNTs in the deposit was approximately 2% by weight by using long CNTs (length 5-9 μm, diameter 110-170 nm) or short CNTs (length < 1 μm, diameter 9.5 nm) in the bath. The density of CNTs is close to 1.3 gm/cm³ and pure Cu is 8.89 gm/cm³ which indicates the CNTs in the deposit are up to 12% by volume.

5. Results

To utilise CNTs as a composite with Cu for interconnect applications it is necessary to verify the influence of the materials on the Cu plating chemistry. Fig. 5 shows the comparison of cyclic voltammetry of Cu and Cu/CNTs co-deposition from a simple $CuSO_4/H_2SO_4$ bath (hereafter referred to as the basic bath) with/without CNTs and surfactant at a scan rate of 0.01 V/s. It can be seen that the addition of CTAB and CNT results in a cathodic peak potential shift to a more negative value which represents a suppression influence on Cu deposition. On the other hand the Cu deposition occurs at lower overpotential when the bath contains either nafion or SDS which represents an acceleration influence. The diffusion coefficient for Cu^{2+} ions estimated from chronoamperometry data using the Cottrel equation in the basic bath (0.24M $CuSO_4$ + 1.8M H_2SO_4) is 4.5 x 10^{-6} cm²/s. A similar value (4.6 x 10^{-6} cm²/s) was found from the SDS containing Cu/CNTs bath. Upon addition of nafion or CTAB in the Cu/CNTs bath, the diffusion coefficient value of Cu^{2+} ions slightly increases to 5.1 x 10^{-6} cm²/s and 5.3 x 10^{-6} cm²/s, respectively. It is clear that CNTs and surfactants in the Cu bath do not have a significant influence on the diffusion coefficient value of Cu. These results indicate that the CNT + surfactant is compatible with the basic Cu sulphate/sulphuric acid bath chemistry. An assessment of the influence of the composite materials on baths that contain the basic constituents and the necessary additives to achieve bottom-up fill or superfilling of interconnect features in silicon technology is also required.

The kinetics of the metal nucleation and growth/dissolution can be analysed with a rotating disk electrode system. While acknowleding the limitations and complications in the kinetic analysis of Cu [12] the general trends indicated in the data are consistent with published data for the Cu sulphate system and those with typical damascene additives. It can be seen from the kinetic data analysis below that the exchange current density, i_0, for Cu nucleation and growth from the basic $CuSO_4$ bath without any additive is 7.24 mA/cm^2 and the E_0 value is - 406 mV. The exchange current value for Cu nucleation and growth in the literature varies from 1 to 15 mAcm^{-2} [73–75]. Addition of all typical additives in $CuSO_4$ bath decreases the exchange current density, i_0 for Cu nucleation and growth from 7.24 mA/cm^2 to 1.2 mA/cm^2 and increases the E_0 value from - 406 mV to - 417.5 mV. This result confirms that all additives together have a suppressor effect on Cu deposition. It can be observed that addition of 1% nafion also has a minor suppressor type behaviour on Cu nucleation and growth as it slightly decreases the exchange current density, i_0 from 7.24 mA/cm^2 to 7.07 mA/cm^2 and increases E_0 value from - 406 mV to - 410.5 mV. But addition of CNTs has an accelerator influence on Cu nucleation and growth increasing the exchange current density, i_0 from 7.24 mA/cm^2 to 10.23 mA/cm^2 and decreasing the E_0 value from - 406 mV to - 403.5 mV. It is also observed that all typical additives including nafion and nanotubes together in the solution have an overall suppressor effect on Cu deposition. The summary results are shown in table 7. It can be seen from the table that anodic slopes are in the range from 1/65 mV to 1/76 mV and cathodic slopes are in the range from 1/122 mV to 1/164 which are close to the theoretical values when the reactions are two separate single electron transfer steps. The above results show that baths containing nafion & CNTs are compatible with the existing typical $CuSO_4$ bath used in IC interconnect deposition.

Conditions	1/slope, mV		E_0/mV	I_0/ mAcm^{-2}
	Anode	Cathode		
0.24 M $CuSO_4$ + 1.8 M H_2SO_4 (Basic bath)	154	67	- 406.0	**7.24**
Basic bath + 1% nafion	161	66	- 410.5	**7.07**
Basic bath + 1% nafion + 10 ppm CNTs	164	76	- 403.5	**10.23**
Basic bath + 50 ppm Cl$^-$ + 300 ppm PEG + 1 ppm SPS	151	67	- 417.5	**0.54**
Basic bath + All additives + 1% nafion	128	72	- 420.0	**1.70**
Basic bath + All additives + 1% nafion + 10 ppm CNTs	122	76	- 421.0	1.66

Table 7. Summary of Tafel analysis obtained from rotating disk system.

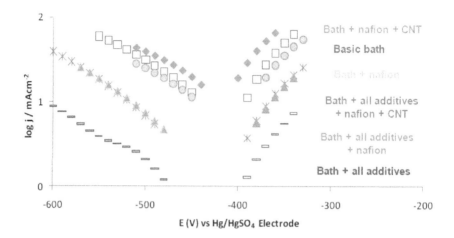

Figure 6. Comparison of Tafel plots for Cu deposition from with/without additives and CNTs in a standard Cu bath using a Cu rotating disk electrode which was rotated at 2000 rpm during experiments (Initial potential: 0 V, scan rate 0.1 V/s). Area of Cu RDE was 12.566 mm²

On a 1 cm² Cu substrate (200 nm sputtered Cu on Si) the deposition current was 15 mA/cm² and deposition time was 1 h. The concentration of CNTs in the bath was 100 mg/l. The acid solution was then vacuum filtered using 5 μm PTFE filter paper. The length and diameter of the MWCNTs were 5-9 μm and 110-170 nm respectively. Fig. 7 shows the SEM images of PTFE membrane after filtration of Cu/CNT deposits. The CNTs are clearly observed.

Figure 7. SEM image of CNTs on PTFE membrane after filtration of dissolved Cu/CNT composites. Image magnification 5000X on left and 40000X on right.

The amounts of CNTs in the deposit are shown in table 8 which compares the percentage of CNTs in the deposit when long or short CNTs with different surfactants were added in the bath. We found the percentage of CNTs was slightly higher (CNT content 1.69% by weight) when SDS was added with long CNTs in the bath (CNT content 1.13% by weight when

CTAB was added). On the other hand, the percentage of CNTs was slightly higher (CNT content 1.64% by weight) when SDS or CTAB was used in the bath with short CNTs in the bath (CNT content 1.12% by weight when nafion was added). It can be seen from the table, the weight percentage of MWCNTs in the deposit was less than 2%. According to the literature [25, 60] the maximum CNT concentration in Cu/CNT composites achieved has been approximately 2.5 % by weight. So the value found in these experiments is quite reasonable. The density of MWCNTs is close to 1.3 g/cm^3 and pure Cu is 8.89 g/cm^3 which indicates that the CNT content in the deposit is up to 12% by volume.

MWCNTs Length μm	MWCNTs Diameter nm	Surfactant	% CNTs
5 – 9	110 - 170	Nafion	1.12
5 – 9	110 - 170	CTAB	1.64
5 – 9	110 - 170	SDS	1.64
<1	9.5	Nafion	1.56
<1	9.5	CTAB	1.13
<1	9.5	SDS	1.69

Table 8. The weight percentage comparison of CNTs in the composite films using different size of CNTs and surfactants in the bath. PTFE membrane was used for filtration purpose.

The electrical properties of Cu/CNT composites were assessed by determining the resistivity of submicron films. Room temperature self-anneal phenomenon is usually observed in electrodeposited Cu films [76-80]. Due to large grain growth at room temperature and annihilation of the defects (void, vacancy, stacking fault, impurities redistribution etc), the electrical resistivity of Cu may change with time. It is therefore necessary to monitor the resistivity changes of Cu and Cu/CNT composite films over time after electrodeposition. The resistivity of Cu and Cu/CNTs composite films at room temperature was monitored using a four point probe apparatus (Keithley 2400 four point probe). In each case we took 4 samples and recorded the average resistivity. The film was electrodeposited on a sputter Cu coated Si substrate. The deposition current density was 15 mAcm^{-2} and deposition time was 2 minutes. The thickness of film was measured by using surface profilometry to be approximately 660 nm.

The electrical resistivity results showed that at room temperature the resistivity of Cu/CNTs composite films (2.46 μΩ-cm) when nafion was used for the surfactant of CNTs is close to the resistivity of Cu film deposited (2.15 μΩ-cm). The resistivity of Cu/CNTs composite film was higher when SDS (3.03 μΩ-cm) or CTAB (4.19 μΩ-cm) was used as a surfactant. The results are summarised in table 9. There was a larger scatter in the distribution of resistivity data in the CTAB case. This is probably a result of a less uniform and void rich deposit from the CTAB containing bath which significantly increased the resistivity.

The resistivity of samples maintained at room temperature did not change significantly. The summary of the changes of the room temperature resistivity over time for Cu and Cu/CNT composite films deposited from different surfactant containing baths are shown in table 10. Osaka et al [66] reported that the resistivity of a deposit from an additive free bath and Cl + PEG containing bath was unchanged with time. But when SPS was present in the bath, the resistivity decreased over time due to self-annealing. Lee and Park [82] reported that self-annealing is caused by Cu grain boundary diffusion. They mentioned that locally high stress originated from the trapped large molecule PEG which can accelerate grain boundary diffusion of Cu. There is a lack of consensus about the cause of self-annealing [81-86]. Among the suggested possible causes for self-annealing of electrodeposited Cu film are bath compositions [83], additives [77, 81, 82], film thickness [79, 80, 84], barrier layers [86, 87], impurities [79, 85] and deposition current [80].

Bath	Resistivity / $\mu\Omega$-cm
Basic (0.24 M CuSO$_4$ + 1.8 M H$_2$SO$_4$)	2.17
Basic + Nafion + CNT	2.43
Basic + SDS + CNT	3.03
Basic + CTAB + CNT	4.69

Table 9. Comparison of the resistivity of Cu and Cu/CNT composite film at room temperature 1 hour after deposition using different surfactants in the bath.

Bath	Time after deposition / hour	Resistivity / $\mu\Omega$-cm
Basic	1	2.17
	312	2.15
Basic + Nafion + CNT	1	2.43
	311	2.47
Basic + CTAB + CNT	1	4.09
	313	4.19

Table 10. Comparison of the resistivity changes of Cu and Cu/CNT composite film at room temperature and 311 to 313 hours after deposition.

It is well known that through annealing at higher temperature a reduced defect Cu microstructure can be obtained [76-87]. Cu/CNT composite films (660 nm in thickness) were annealed in nitrogen at 215°C, 265°C and 315°C for 20 minutes. It can be seen that a clear decrease of sample resistivity was observed with increasing annealing temperature which is shown in table 11. The resistivity value of Cu film approaches that of bulk Cu value (1.67 $\mu\Omega$-cm) after annealing at 315°C for 20 minutes. Also the resistivity of Cu/CNTs composite films decreased with increasing annealing temperature. The electrical resistivity of the Cu/CNTs composite films deposited from a nafion and CTAB containing bath became 1.88 $\mu\Omega$-

cm and 2.10 µΩ-cm respectively when the sample was annealed at 315 °C. The conductivity increase of the composite films was probably due to a decrease the interface resistance between CNTs and Cu matrix at the higher temperature, grain refinement and elimination of defects under high temperature annealing.

Annealing temperature °C	Basic bath	Resistivity / µΩ-cm Basic + Nafion + CNT	Basic + CTAB + CNT
No anneal	2.15	2.46	4.19
215	1.78	2.14	2.72
265	1.92	2.04	2.45
315	1.67	1.88	2.10

Table 11. Comparison of the resistivity of Cu and Cu/CNT composite films at room temperature and higher temperatures 312 hours after deposition using different surfactants in the bath.

The influence of CNT concentration in the Cu/CNT composite bath was investigated. The electrical resistivity results showed (fig. 8) that at room temperature the resistivity increased 10% when the concentration of CNT in the bath was increased from 10 mg/l to 100 mg/l. This data also shows no evidence of self annealing at room temperature for the composite material.

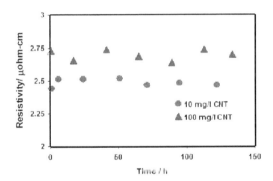

Figure 8. Comparison of the resistivity of Cu/CNT composite film at room temperature over time using 10 mg/l and 100 mg/l CNT in the bath.

It can be seen from fig. 9 that when the samples were annealed at higher temperature up to 315ºC for 20 minutes the resistivity decreased from 2.46 µΩ-cm to 1.89 µΩ-cm for 10 mg/l CNT and 2.7 µΩ-cm to 2.19 µΩ-cm for 100 mg/l CNT in the bath. It is expected that CNT content in the composite is higher when deposited from higher CNT concentration containing bath [25, 60]. The resistivity increase of the higher CNT content bath is probably due to increased CNTs content and higher contact resistance between CNTs and Cu in the composites.

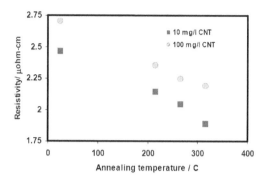

Figure 9. Comparison of the resistivity change of Cu/CNT composite film after annealing for 20 minutes at higher temperature using 10 mg/l and 100 mg/l CNT in the bath.

In summary, the Cu/CNT composites were codeposited with the aid of a surfactants at different current densities. As a comparative study, three surfactants (nafion, CTAB and SDS) were used separately to disperse CNTs in the bath and Cu/CNT composite films were electrodeposited. SDS in the bath results in a smoother deposition whereas CTAB leads to rougher deposition of the composite. The maximum CNT concentration in Cu/CNT composites achieved in our study was approximately 2 % by weight deposited from 100 mg/l CNT containing composite baths. The electrical resistivity results show that at room temperature the resistivity of Cu/MWCNT composite film (2.47 $\mu\Omega$-cm) is close to the resistivity of Cu film (2.15 $\mu\Omega$-cm) when nafion was used in the deposition bath for the surfactant of MWCNTs. With the use of CTAB or SDS in the bath, the resistivity of Cu/MWCNT film was higher [deposited from 10 mg/l CNT containing composite baths]. A clear decrease in sample resistivity was observed with increasing annealing temperature. The resistivity also increased when the concentration of MWCNTs was increased from 10 mg/l to 100 mg/l in the bath.

The line resistance of Cu filled and Cu/CNT filled test chip structures was measured using a Cascade probe station. The test chip structure consisted of 110 μm x 80 μm pads connected with metal lines of different widths. A Cu seed (12 nm) and a barrier Ta/TaN (25 nm) were PVD deposited. To achieve a uniform deposit the plated substrates were planarised with a CMP process. The test chip coupon was mounted on a 4 inch Si carrier-wafer. A Logitech CDP51 was used for the CMP. The polishing slurry used was a Cabot Microelectronics product, Eterpol 2362, which was mixed with H_2O_2 (30%), the ratio of H_2O_2 to slurry was 5% by volume. During CMP, the rotation of wafer holder and polishing pad was 50 rpm and the applied pressure was 2-3 psi. Fig. 10 shows the SEM image of the test structure after CMP for 2 minutes. It can be seen that excess deposits were completely removed by the developed CMP process. Before probing, the test sample was vacuum attached in a dedicated holder. Four micro-probes were placed on four pads in the structure. The pads were connected with Cu filled interconnect lines. The Cu was deposited from a damascene additive containing sulphate based bath. Fig. 11 shows the electrical measurement of 110 nm width Cu line in the test structure. It can be seen from the measurement that the I-V curves of Cu deposited line was linear and the resistance value was estimated to be 284 Ω which is to be expected for lines of that dimension.

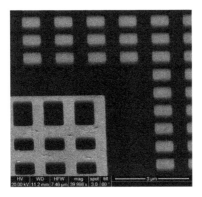

Figure 10. Plan view SEM image of test structure after complete CMP. The structure was filled by electrodeposited Cu (lighter colour in image). Image magnification 40000X.

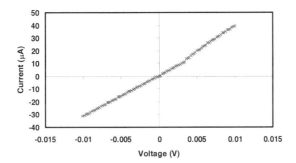

Figure 11. Current vs. voltage curve of the 110 nm line width connected with four 110 μm x 80 μm pads. The features were filled with electrodeposited Cu.

The next samples investigated contained CNTs in the Cu deposited from the nafion containing bath. The concentrations of CNTs and nafion were 50 mg/l and 0.5% respectively. Fig. 12 shows the electrical measurement of 110 nm line width filled with Cu/MWCNT in the test structure. It can be seen from the measurement that the I-V curve of the Cu/MWCNT deposited line was linear and the resistance value was 29.7 kΩ. The resistance of individual MWCNT is hundreds of kΩ (minimum resistance for a ballistic single-walled CNT is ~ 6.5 kΩ). The high resistance of individual CNTs is due to high contact and quantum resistance. Therefore, relatively dense arrays of CNTs will be needed to replace Cu interconnects. As Cu and MWCNTs were codeposited in the narrow line with 110 nm width, so the resistance in Cu/MWCNT composites is expected to be between the resistance value of Cu (284 Ω) and MWCNT (hundreds kΩ). The resistance of the line filled with Cu/CNTs could be improved by using SWCNT instead of the MWCNTs used in this composite.

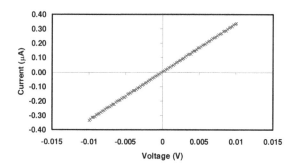

Figure 12. Current vs. voltage curve of the 110 nm line width which was filled with electrodeposited Cu/MWCNT composite. The concentrations of nafion and MWCNTs in the bath were 0.5% and 50 mg/l respectively.

6. Conclusion

In this chapter we have reported the influence of surfactants on the properties of Cu/CNT composites on Si substrates. Cu/CNTs composite films were co-deposited by electrodeposition. Before electrodeposition, CNTs were dispersed by a suitable surfactant. Electrochemical data shows that nafion or SDS accelerates the co-deposition where as CTAB suppresses the deposition. Nafion and SDS surfactants result in a relatively smooth deposit whereas CTAB surfactant leads to rougher deposition of the composite. The amount of CNTs in the deposit was up to 2 % by weight using different surfactants and different length/diameter of CNTs. Our electrical analysis showed that for Cu/CNT composite samples maintained at room temperature, the resistivity over time did not change significantly. The electrical resistivity results also showed that at room temperature the resistivity of the Cu/CNT composites film (2.43 μΩ cm) is close to the resistivity of Cu film (2.17 μΩ cm) when nafion was used in the bath to disperse the CNTs. The resistivity of Cu/CNTs film was higher when CTAB or SDS were used instead of nafion as a surfactant. The electrical resistivity results showed that at room temperature the resistivity increased 10% when the concentration of CNT in the bath was increased from 10 mg/l to 100 mg/l. A clear decrease of sample resistivity of composite films was observed with increasing annealing temperature. Cu/CNT composites deposited at a test structure with submicron lines and vias with Cu/CNT composites was only possible from the nafion surfactant containing damascene. The electrical measurement of 110 nm line width filled with Cu/MWCNT showed that the I-V curves of the Cu/MWCNT deposited line was linear and the resistance value was 29.7 kΩ which was significantly higher that the resistance value of Cu (284 Ω) deposited. Improvements on these values will require lower resistance SWCNTs or the improvement of the density of aligned nanotubes in the composite structure. This may be more feasible in larger dimension features such as those required for TSV interconnect at the chip scale rather the use of composites for IC interconnect at deep sub micron dimensions.

Acknowledgements

This research was supported by the Irish Research Council for Science, Engineering and Technology (IRCSET) postgraduate scholarship Enterprise Partnership scheme in collaboration with Intel Ireland Ltd., funded under the National Development Plan.

Author details

Tamjid Chowdhury and James F. Rohan

Tyndall National Institute, University College Cork, Lee Maltings, Cork, Ireland

References

[1] L. L. Vadasz, A. S. Grove, T. A. Rowe and G. E. Moore, Silicon gate technology. IEEE Spectrum 6, 28 (1969).

[2] R. Solanki and B. Pathangey, Electrochem. Solid St., 3, 479 (2000).

[3] S. Venkatesan, A. Gelatos, S. Hisra, B. Smith, R. Islam, J. Cope, B. Wilson, D. Tuttle, R. Cardwell, S. Anderson, M. Angyal, R. Bajaj, C. Capasso, P. Crabtree, S. Das, J. Farkas, S. Filipiak, B. Fiordalice, M. Freeman, P. Gilbert, M. Herrick, A. Jain, H. Kawasaki, C. King, J. Klein, T. Lii, K. Reid, T. Saaranen, C. Simpson, T. Sparks, P. Tsui, R. Venkatraman, D. Watts, E. Weitzman, R. Woodruff, I. Yang, N. Bhat, G. Hamilton and Y. Yu, Proc. IEEE-IEDM, 97, 769 (1997).

[4] D. Edelstein, J. Heidenreich, R. Goldblatt, W. Cote, C. Uzoh, N. Lustig, P. Roper, T. McDevitt, W. Motsiff, A. Simon, J. Dukovic, R. Wachnik, H. Rathore, R. Schulz, L .Su, S. Luce and J. Slattery, Proc. IEEE-IEDM, 97, 773 (1997).

[5] M. T. Wang, Y. C. Lin and M. C. Chen, J. Electrochem. Soc., 145 (7), 2538 (1998).

[6] C. K. Hu, B. Luther, F. B. Kaufman, J. Hummel, C. Uzah and D. J. Pearson, Thin Solid Films, 262, 84 (1995).

[7] K. Abe, Y. Harada and H. Onoda, Proc. 13th Intern VLSI Multilevel Interconnect Conf. 308 (1995).

[8]   J. S. H. Cho, H. K. Kang, I. Asano and S. S. Wang. IEDM Tech Digest 297 (1992).

[9]   P. C. Andricacos, C. Uzoh, J. O. Dukovic, J. Horkans and H. Deligianni, IBM J. Res. Dev., 42, 567 (1998).

[10] K. Kondo, K. Hayashi, Z. Tanaka and N. Yamakawa, ECS Proceedings on Electrochemical Processing in ULSI Fabrication, 8, 76 (2000).

[11] J. Kelly and A. West, Electrochem. Solid-St., 2, 561 (1999).

[12] J. P. Healy, D. Pletcher and M. Goodenough, J. Electroanal. Chem., 338, 179 (1992).

[13] C. K. Hu and J. M. E. Harper, Mater. Chem. Phys., 52, 5 (1998).

[14] B. Li, T. Sullivan, T. Lee and D. Badami, Microelectron. Reliab., 44. 365 (2004).

[15] C. Hu, L. Gignac and R. Rosenberg, Microelectron. Reliab., 46. 213 (2006).

[16] C. Hu, L. Gignac and R. Rosenberg, E. Liniger, J. Rubino and C. Sambucetti, Microelectron. Eng., 70, 406 (2003).

[17] J. Gambino, J. Wynne, J. Gill, S. Mongeon, D. Meatyard, B. Lee, H. Bamnolker, L. Hall, N. Li, M. Hernandez, P. Little, M. Hamed, I. Ivanov and C. Gan, Microelectron. Eng., 83, 2059 (2006).

[18] S. Iijima, Nature, 354, 56 (1991).

[19] H. Dai, A. Javery, E. Pop, D. Mann and Y. Lu, Nano: Brief Reports and Reviers, 1, 1 (2006).

[20] N. Srivastava and K. Banerjee, Proc 21st Int Multilevel Interconnect, 393 (2004).

[21] P. Rapposelli, B. Capraro, J. Dijon, G. Groeseneken, D. Cott, J. Pinson, X. Joyeux, J. Amadou, J. Noyen and B. Sels, ECS Trans., 25 (10), 63 (2009).

[22] International Technology Roadmap for Semiconductors, 2011 update, http://www.itrs.net/Links/2011Winter/11_Interconnect.pdf (accessed 13 July 2012)

[23] Y Chai, K. Zhang, M. Zhang, P. Chen and M. Yuen., Electronic Components and Technology Conference, IEEE, 1224 (2007).

[24] Y Chai, P. Chan and Y. Fu., Electronic Components and Technology Conference, IEEE, 412 (2008).

[25] J. Yoo, J. Song, J. Yu, H. Lyeo, S. Lee and J. Hahn., Electronic Components and Technology Conference, ECTC, 1282 (2008).

[26] Q. Chen, G. Chai and Bo Li., Proc. IMechE, 219, 67 (2006).

[27] O. Hjorstam, P. Isberg, S. Soderholm and H. Dai., Appl. Phys. A., 78, 1175 (2004).

[28] P. Liu, D. Xu, Z. Li, B. Zhao, E. Kong and Y. Zhang, Microelectron. Eng., 85, 1984 (2008)

[29] G. Chai, Y. Sun, J. Sun and Q. Chen., J. Micromech. Microeng., 18, 35013 (2008)

[30] Q. Chan, US patent application 11/437,180, filed May, 2006.

[31] Y. Son, J. Yoo and J. Yu, US patent application 11/589,305, filed Oct, 2006.

[32] P. Lo, J. Wei, B. Chen, J. Chiang and M. Kao, US patent application 11/289,523, filed Dec, 2005.

[33] V. Davis, A. Parra-Vasquez, M. Green, P. Rai, N. Behabtu, V. Prieto, R. Booker, J. Schmidt, E. Kesselman, W. Zhou, H. Fan, W. Adams, W. Hauge, J. Fischer, Y. Cohen, Y. Talmon, R. Smalley and M. Pasquali, Nat. Nanotechnol., 4, 830 (2009).

[34] F. Pompeo and D. Resasco, Nano Lett., 2, 369 (2002).

[35] L. Feng, H. Li, F. Li, Z. Shi and Z. Gu, Carbon, 41, 2385 (2003).

[36] Y. Lin, F. Allard and Y. Sun, J. Phys. Chem. B, 108, 3760 (2004).

[37] K. Matsuura, K. Hayashi and N. Kimizuka, Chem. Lett., 32, 212 (2003).

[38] W. Huang, S. Taylor, K. Fu, Y. Lin, D. Zhang, T. Hanks, A. Rao and Y. Sun, Nano Lett., 32, 212 (2003).

[39] H. Peng, L. Alemany, J. Margrave and V. Khabashesku, J. Am. Chem. Soc., 125. 15174 (2003).

[40] Y. Wang, Z. Iqbal, S. Mitra, J. Am. Chem. Soc., 128, 95 (2006).

[41] D. W. Schaefer, J. M. Brown, D. P. Anderson, J. Zhao, K. Chokalingam, D. Tomlin and J. Ilavsky, J. Appl. Crystallogr. 36, 553 (2003).

[42] J. Li, Q. Ye, A. Cassell, H. Tee Ng, R. Stevens, J. Han and M. Meyyappan, Appl. Phys. Lett., 82, 2491(2003).

[43] J. Lee, U. Paik, J. Choi, K. Kim, S. Yoon, J. Lee, B. Kim, J. Kim, M. Park, C. Yang, K. An, Y. Lee, J. Phys. Chem. C, 111, 2477 (2007).

[44] F. Ko, C. Lee, C. Ko and T. Chu, Carbon, 43, 727 (2005).

[45] Y. Chen and Y. Mitra, J. Nanosci. Nanotechnol.., 11, 5770 (2008).

[46] J. Li and H. Greenburg, Chem. Eur. J, 12, 3869 (2006).

[47] C. Lau, R. Cervini, S. Clarke, M. Markovic, J. Matisons, S. Hawkins, C. Huynh and G. Simon, J. Nanopart Res, 10, 77 (2008).

[48] S. Agarwal, M. Raghuveer, H. Li and G. Ramanath, Appl. Phys. Lett., 90, 193104 (2007).

[49] H. Tantang, J. Ong, C. Loh, X. Dong, P. Chen, Y. Chen, X. Hu, L. Tan and L. Li, Carbon, 47, 1467 (2009).

[50] M. O'Connell, P. Boul, L. Ericson, C. Huffman, Y. Wang, E. Haroz, C. Kuper, J. Tour, K. Ausman and R. Smalley, Chem. Phys. Lett., 342, 265 (2001)

[51] X. Chen, C. Cheng, H. Xiao, H. Liu, L. Zhou, S. Li and G. Zhang, Tribol. Int., 39, 22 (2006).

[52] X. Chen, F. Cheng, S. Li, L. Zhou and D. Li, Surf. Coat. Technol., 155, 274 (2002).

[53] C. Guo, Y Zuo, X. Zhao, J. Zhao and J. Xiong, Surf. Coat. Technol., 202, 3385 (2008)

[54] Y.Yang, Y. Wang, Y. Ren, C. He, J. Deng, J. Nan, J. Chen and L. Zuo. Mat. Lett., 62, 47 (2008)

[55] B. Praveen, T. Venkatesha, Y. Naik and K. Prashantha, Surf. Coat. Technol., 201, 5836 (2007)

[56] B. I. Yakobson and R.E. Smalley, Am. Sci, 85, 324 (1997).

[57] L. Xu, X. Chen, W. Pan, W. Li, Z. Yang and Y. Pu., Nanotechnol, 18, 435607 (2007)

[58] X. Chen, C. Cheng, H. Xiao, H. Liu, L. Zhou, S. Li and G. Zhang, Tribol. Int., 39, 22 (2006).

[59] N. Tuerui, B. Fugetsu and S. Tanaka, Anal. Sci., 22, 895 (2006).

[60] S. Arai, M. Endo, T. Sato and A. Koide, Electrochem. Solid-St., 9, C131 (2006)

[61] L. Shi, C. Sun, P. Gao, F. Zhou and W. Liu., Surf. Coat. Technol, 200, 4870 (2006)

[62] Y Tsai, S. Li and J. Chen, Proc IEEE Conf on Nanotechnol, (2005)

[63] C. Engtrakul, M. Davis, T. Gennett, A. Dillon, K Jones and M. Heben, J. Am. Chem. Soc., 127, 17548 (2005)

[64] S. Arai, A. Fujimori, M. Murai and M. Endo, Mat. Lett., 62, 3545 (2008).

[65] S. Arai, T. Saito and M. Endo, Electrochem. Solid-St., 11, D72 (2008).

[66] T. Osaka, N. Yamachika, M. Yoshino, M. Hasegawa, Y. Negishi and Y. Okinaka, Electrochem Solid-St., 12, D15 (2009).

[67] P. Dixit, C Tan, L. Xu, N. Lin, J. Miao, J. Pang, P. Backus and R. Preisser, J. Micromech. Microeng., 17, 1078 (2007).

[68] A. Yeoh, M. Chang, C. Pelto, T. Huang, S. Balakrishnan, G. Leatherman, S. Agraharam, W. Guota, Z. Wang, D. Chiang, P. Stover and P. Brandenburger, IEEE 56th Electronic Components and Technology Conference, 1611 (2006).

[69] A. Braun, Semiconductor International, 3D Integration in Design and Test Support. Article ID CA6615469.

[70] O. Luhn, A. Radisic, P. M. Vereecken, C. van Hoof, W. Ruythooren and J. P. Celis, Electrochem. Solid-St., 12 (5), D39 (2009).

[71] R. Beica, C. Sharbono and T. Ritzdorf, DTIP of MEMS & MOEMS, ISBN: 978-2-355500-006-5, 2008.

[72] A. Radisic, O. Lühn, J. Vaes, S. Armini, Z. Mekki, D. Radisic, W. Ruythooren, and P. Vereecken, ECS Trans., 25 (38), 119 (2010).

[73] E. Mattsson and J. O'M Bockris., Trans. Faraday. Soc., 55, 1586 (1958)

[74] C. H. Yang, Y. Y. Wang and C. C. Wan, J. Electrochem. Soc., 146 (12), 4473 (1999).

[75] S. Varvara, L. Muresan, I. C. Popescu and G. Maurin, J. Appl. Electrochem., 33, 685 (2003).

[76] K. Pantleoan and M. A. J. Somers, J. Appl Phys., 100, 114319 (2006).

[77] V. A. Vasko, I. Tabakovic, S. C. Riemer and M. T. Kief, Electrochem., Solid-St., 6, 100 (2003).

[78] P. M. Vereecken, R. A. Binstead, H. Deligianni and P. C. Andricacos, IBM J. Res & Dev., 49, 3 (2005).

[79] S. Lagrange, S. H. Brongersma, M. Judelewicz, A. Saerens, I. Vervoort, E. Richard, R. Palmans and K. Maex, Microelectron. Eng., 50, 449 (2000).

[80] J. M. E. Harper, C. Cabral, P. C. Andricacos, L. Gignac, I. C. Noyan, K. P. Rodbell and C. K. Hu, J. Appl. Phys., 86, 2516 (1999).

[81] T. Osaka, N. Yamchika, M. Yoshino, M. Hasegawa, Y. Neigishi and Y. Okinaka, Electrochem. Solid-St., 12, D15 (2009).

[82] C. Lee and C. Park, Jpn. J. Appl. Phys., 42, 4484 (2003).

[83] T. Hara, H. Toida and Y. Shimura, Electrochem. Solid-St., 6, G98 (2003) 204

[84] W. H. Teh, L. T. Kon, S. M. Chen, J. Xie, C. Y. Li and P. D. Foo, Microelectron J., 32, 579 (2001).

[85] S. H. Brongersma, E. Richard, I. Vervoot, H. Bender, W. Vandervost, S. Lagrange, G. Beyer and K. Maex, J. Appl. Phys., 86, 3642 (1999).

[86] S. Balakumar, R. Kumar, Y. Shimura, K. Namiki, M. Fujimoto, H. Toida, M. Uchida and T. Hara, Electrochem. Solid-St., 7, G68 (2004).

[87] H. Lee, S. S. Wong and S. D. Lopatin, J. Appl. Phys., 93, 3796 (2003).

Carbon Nanotubes for Green Technologies

Carbon Nanotubes Influence on Spectral, Photoconductive, Photorefractive and Dynamic Properties of the Optical Materials

Natalia V. Kamanina

Additional information is available at the end of the chapter

1. Introduction

It is well known that optoelectronics, telecommunication systems, aerospace, and correction of amplitude-phase aberration schemes, as well as laser, display, solar energy, gas storage and biomedicine techniques are searching for the new optical materials and for the new methods to optimize their properties. So many scientific and research groups are involving in this process and are opening the wide aspects of different applications of new materials, especially optical ones. It has been going on last century that simple manufacturing, design, ecology points of view, etc. indicate good advantage of the nanostructured materials with improved photorefractive parameters among other organic and inorganic systems.

Really, it should be tell that photorefractive properties change is correlated with the spectral, photoconductive and dynamics ones. The change in nonlinear refraction and cubic nonlinearity reveals the modification of barrier free electron pathway and dipole polarizability. From one side it is connected with the change of the dipole moment and the charge carrier mobility, from other side, it is regarded to the change of absorption cross section. Thus, this feature shows the unique place of photorefractive characteristics among other ones in order to characterize the spectral, photoconductive, photorefractive and dynamic properties of the optical materials.

It should be mentioned that promising nanoobjects, such as the fullerenes, the carbon nanotubes (CNTs), the quantum dots (QDs), the shungites, and the graphenes permit to found different area of applications of these nanoobjects [1-6]. The main reason to use the fullerenes, shungites, and quantum dots is connected with their unique energy levels and high value of electron affinity energy. The basic features of carbon nanotubes and graphenes are

regarded to their high conductivity, strong hardness of their C-C bonds as well as complicated and unique mechanisms of charge carrier moving.

These peculiarities of carbon nanoobjects and their possible optoelectronics, solar energy,
gas storage, medicine, display and biology applications connecting with dramatic improvement of photorefractive, spectral, photoconductive and dynamic parameters will be under
consideration in this paper. In comparison with other effective nanoobjects the mail accent
will be given namely on carbon nanotubes (CNTs) and their unique features to modify the
properties of the optical materials.

2. Experiment

The different experimental techniques have been used to study the properties of nanostructured materials.

To reveal the change of the photorefractive properties, as the systems under study the organic thin films based on conjugated monomer, polymer and liquid crystals sensitized with
carbon nanotubes, fullerenes, shungites, graphenes oxides, or quantum dots have been chos-
en. Polyimides (PI), 2-cyclooctylamino-5-nitropyridine (COANP), N-(4-nitrophenyl)-(L)-pro-
linol (NPP), 2-(N-prolinol)-5-nitropyridine (PNP), nematic liquid crystals (NLCs) have been
considered as organic matrixes. These conjugated systems are the good model with effective
intramolecular charge transfer process which can be easy modify via sensitization by nanoobjects. Carbon nanotubes, fullerenes, shungites, graphenes oxides, quantum dots content
was varied in the range of 0.003-5.0 wt.%. The solid thin films have been developed using
centrifuge deposition. The general view of these films is shown in Fig.1. The thickness of the
films was 2-5 micrometers. The LC cell thickness was 5-10 micrometers.

Figure 1. Photographs of samples of pure (nk8) and nanoobjects-containing (nk10) PI films.

The nanostructured LC films have been placed onto glass substrates covered with transparent conducting layers based on ITO contacts. The nanostructured monomer or polymer sol-

id films have been deposited on the substrate with ITO contact. For the electric measurements of volt-ampere parameters, gold contact has been put to the solid thin films upper side. The picture which can interpret the placement of the conducting contacts on the solid conjugated organic thin films is shown in Fig.2.

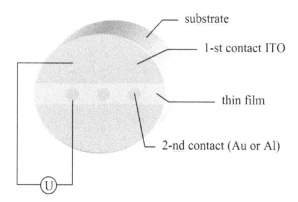

Figure 2. Interpretation of the solid thin films with the conducting layers

The bias voltage applied to the photosensitive polymer layers has been varied from 0 to 50 V. The current–voltage characteristics have been measured under the illumination conditions from dark to light. Voltmeter-electrometer B7−30 and Characteriscope−Z, type TR-4805 has been used for these photoconductive experiments.

The photorefractive characteristics have been studied using four-wave mixing technique analogous to paper [7]. The experimental scheme is shown in Fig.3.

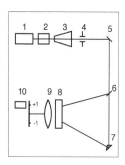

Figure 3. An experimental scheme: 1 – Nd-laser; 2 – second harmonic convertor; 3 – telescope; 4 – diaphragm; 5 – rotating mirror; 6 – beam-splitting mirror; 7 – prism; 8 – sample; 9 – lens; 10 - photodetector.

The second harmonic of pulsed Nd-laser at wave length of 532 nm has been used. The laser energy density has been chosen in the range of 0.005-0.9 J×cm⁻². The nanosecond laser regime with the pulse width of 10-20 ns has been applied. The amplitude-phase thin gratings have been recorded under Raman-Nath diffraction conditions according to which $\Lambda^{-1} \geq d$, where Λ^{-1} is the inverse spatial frequency of recording (i.e., the period of the recorded grating) and d is the film thickness. In the experiments the spatial frequency was in the range of 90-150 mm⁻¹.

The spectral characteristics have been tested using Perkin Elmer lambda 9 spectrophotometer. Dynamic features of nanoobjects-doped LC films have been studied via the four-wave mixing technique and the Frederick's scheme one. Atomic force microscopy (AFM) method using equipment of "NT-MDT" firm, "Bio47-Smena" in the "share-force" regime has been applied to analyze the diffraction relief into the solid conjugated nanostructured thin film.

3. Results and discussion

It should be noticed that previously, we demonstrated [5,6,8] the formation of barrier-free charge transfer pathways, increased dipole moment, and increased specific (per unit volume) local polarizability in some organic matrices doped with fullerenes, carbon nanotubes and quantum dots, where the formation of intermolecular complexes predominated over the intramolecular donor–acceptor interaction. The possible schemes of charge transfer between matrix organic molecule donor fragment and different efficient nanosensitizers including the additional graphene and shungites nanostructures are schematically shown in Fig. 4.

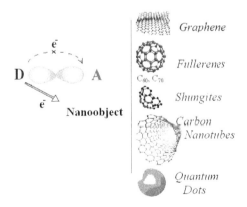

Figure 4. Schematic diagram of possible intermolecular charge transfer domination under intramolecular ones.

Analyzing the Fig.4, one can say that it is necessary to take into account that the charge transfer between matrix organic molecule donor fragment and nanasensitizers can be organized due to their high electron affinity energy (for example, electron affinity energy is close to 2 eV for shungites [9], to 2.65 eV for fullerenes [5,8] and to 3.8-4.2 eV for quantum dots [10]) that is more than the ones for intramolecular acceptor fragments (for example, electron affinity energy of COANP acceptor fragment is close to 0.54 eV [11] and to 1.14-1.4 eV for polyimide one [12]). Regarding graphenes it is necessary to take into account the high surface energy and planarity of the graphenes plane which can provoke to organize the charge transfer complex (CTC) with good advantage too. Regarding the CNTs it should be drawn the attention on the variety of charge transfer pathways, including those along and across a CNT, between CNTs, inside a multiwall CNT, between organic matrix molecules and CNTs, and between the donor and acceptor fragments of an organic matrix molecule.

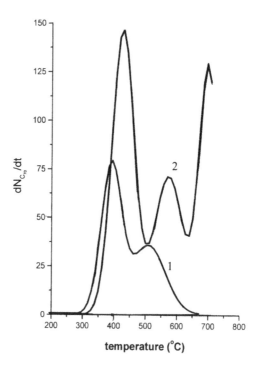

Figure 5. The rate of release of C_{70} molecules on heating of systems: (*1*) COANP with 5 wt % of C_{70} and (*2*) polyimide with 0.5 wt % of C_{70}

It should be noticed that some supporting CTC results for PIs and COANP systems sensitized with nanoobjects can be presented via mass-spectrometry experiments. It is easy to show the organization of CTC using fullerenes acceptor. Really, the mass spectroscopy data point to the effective CTC formation between fullerene and donor part of PI (triphenylamine) and between fullerene and the HN group of COANP systems, respectively. For the 5 wt.% C_{70}-COANP film, mass spectrometry curve contains two peaks. The first one at 400 C corresponds to the release rate of fragments with free fullerene masses. The second one is shifted to the temperature range of 520 °C and associated with the decomposition temperature of the fullerene-HN group complex. For the 0.5 wt.% C_{70}-PI film, curve contains three peaks. The first one is observed also close to 400 C. The second peak is located at 560 °C and associated with the decomposition of fullerene- triphenylamine complex. It should be noticed that the melting temperature of these PIs is 700-1000 °C, thus, the third peak at the temperature higher 700 °C corresponds to the total decomposition of PI. Figure 5 presents the mass-spectrometry data.

By monitoring the diffraction response manifested in the laser scheme (see Fig.6); it is possible to study the dynamics of a photo-induced change in the refractive index of a sample and to calculate via [13] the nonlinear refraction and nonlinear third order optical susceptibility (cubic nonlinearity). An increase in the latter parameter characterizes a change in the specific (per unit volume) local polarizability and, hence, in the macroscopic polarization of the entire system.

Figure 6. The visualization of the diffraction response in the organic films doped with nanoobjects.

The main results of this study are summarized in the Table 1 (Ref. 5,6,9,14-18) in comparison to the data of some previous investigations. An analysis of data presented in the Table 1 for various organic systems shows that the introduction of nanoobjects as active acceptors of electrons significantly influences the charge transfer under conditions where the intermolecular interaction predominates over the intramolecular donor–acceptor ones. Moreover, redistribution of the electron density during the recording of gratings in nanostructured materials changes the refractive index by at least one order of magnitude in comparision to that in the initial matrix. The diffusion of carriers from the bright to dark region during the laser recording of the interference pattern proceeds in three (rather than two) dimensions, which is manifested by a difference in the distribution of diffraction orders along the horizontal and vertical axes (see Fig.6). Thus, the grating displacement takes place in a three dimensional (3D) medium formed as a result of the nanostructirization (rather than in a 2D medium).

Some atomic force microscopy data are supported the realization of 3D-media via development of complicated diffraction relief into the solid thin conjugated films after transfer from the reversible regime to the irreversible one. Figure 7 demonstrates this fact. Two types of diffraction replica, namely due to interference of laser beams onto the thin films surface and due to the diffraction of these beams inside the body of the nanostructured media have been presented.

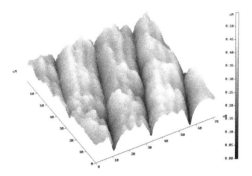

Figure 7. Demonstration of AFM evidence of new 3D-media development.

Using the obtained results, the nonlinear refraction n_2 and nonlinear third order optical susceptibility (cubic nonlinearity) $\chi^{(3)}$ for all systems have been calculated using a method described in [13,18]. In the current experiments using four-wave mixing technique, the nonlinear refraction coefficient and cubic nonlinearity (third order susceptibility) have been estimated via equations (1) and (2):

$$n_2 = \Delta n_i / I \tag{1}$$

$$\chi^{(3)} = n_2 n_0 c / 16\pi^2 \tag{2}$$

where I – is the irradiation intensity, n_o – is the linear refractive index of the media, c – is the speed of the light.

It was found that these parameters fall within $n_2 = 10^{-10}$–10^{-9} cm²/W and $\chi^{(3)} = 10^{-10}$–10^{-9} cm³/erg.

Moreover, it should be remained, that optical susceptibility $\chi^{(n)}$, from fundamental point of view, directly connected with the dipole system polarizability [(n)] via equation (3) written in the paper [19]:

$$\chi^{(n)} = \alpha^{(n)} / \upsilon \tag{3}$$

where $\alpha^{(n)}$ – dipole polarizability and υ – local volume.

Therefore, using the fact that polarizability of all structures can be accumulated from local volumes, it can be found that increased micropolarization of system (see eq.4) will predict the dynamic properties improvement and high electro-optical response speed.

$$P^{\circ} = {}^{\circ}\chi^{(1)}E^{\circ} + {}^{\circ}\chi^{(2)}E^{2\circ} + \chi^{(3)}E^{3\circ} + \ldots + {}^{\circ}\chi^{(n)}E^{n\circ} + \ldots \tag{4}$$

It should be mentioned (see Table 1) that the larger nonlinear optical parameters have been found for CNTs-doped organic systems or CNTs-nanofibers-doped ones. It is natural to suggest (see Fig.8) that variations of the length, of the surface energy, of the angle of nanoobject orientation relative to the intramolecular donor can significantly change the pathway of charge carrier transfer, which will lead to changes in the electric field gradient, dipole moment (proportional to the product of charge and distance), and mobility of charge carriers.

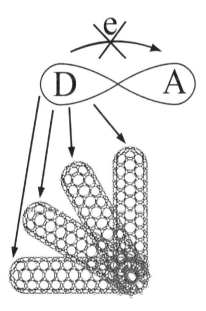

Figure 8. Schematic diagram of possible charge transfer pathways depending on the arrangement of introduced intermolecular acceptor relative to the intramolecular donor

In addition, the barrier free charge transfer will be influenced by competition between the diffusion and drift of carriers during the creation of diffraction patterns with various periods and, hence, differing charge localization at the grating nodes and antinodes. Indeed, in the case of a nanocomposite irradiated at small spatial frequencies (large periods of recorded grating), a drift mechanism of the carrier spreading in the electric field of an intense radiation field will moist probably predominate, while at large spatial frequencies (short periods of recorded grating) the dominating process is diffusion. This also naturally accounts for the aforementioned discrepancy of published data on photoinduced changes in the refractive index of nanocomposites, greater values of which were observed (see, e.g., data presented in the table for systems doped with CNTs and MIG nanofibers) at smaller spatial frequencies. Lower values of photoinduced changes in the refractive index of nanocomposites were observed at high spatial frequencies. This evidence predicts the strong correlations between photorefractive and photoconductive parameters.

To support the evidence on correlation between photorefractive and photoconductive features of the materials studied, the volt-current characteristics for nanoobjects-doped solid thin films and pure ones has been measured. After that charge carrier mobility has been estimated using the Child–Langmuir current–voltage relationship [20] following the formula (5) shown below:

$$\mu = 10^{13} I d^{3} \times \varepsilon^{-1} V^{-1} \tag{5}$$

For example, one can calculate the absolute values of the charge carrier mobility in pure and fullerene-modified PI samples. The results of these calculations show that the introduction of fullerenes leads to a tenfold increase in the mobility. The absolute values were estimated for a bias voltage of 10 V, a film thickness of $d = 2$ μm, a dielectric constant of $\varepsilon \sim 3.3$, a fullerene content of about 0.2 wt % C_{70}, and an upper electrode contact area with a diameter of 2 mm. Under these conditions, the carrier mobility in a fullerene-modified polyimide PI film is ~0.3×10^{-4} cm^2/(V s), while the analogous value for pure PI is ~0.17×10^{-5} cm^2/(V s). These values well agree with the data reported in [21], where it was demonstrated that the carrier mobility in pure PI films ranges in the interval from 10^{-7} to 0.5×10^{-5} cm^2/(V s). Relationship (5) used for the estimation of charge carrier mobility is valid in the case of currents limited by the space charge. This situation is characteristic of most of the conjugated organic structures (in particular, PIs) in which the charge transfer processes are additionally determined by traps, although formula (5) contains no terms dependent on the illumination intensity. However, taking into account the aforementioned equality of the activation energies of conductivity and mobility in PIs, the results of calculations of the relative changes in the carrier mobility probably adequately reflect the general trends in mobility variations. This behavior does not contradict the pattern of changes in the mobility observed for the other conjugated organic systems, for example, for the fullerene–carbazole one [22].

We have also calculated the relative values of the charge carrier mobility μ and estimated that two orders of magnitude differences of charge carrier mobility for the pure and nanoobjects-doped films has been found. Moreover, the following relation for the charge carrier mobility has been proposed:

$$\mu_{\text{pure organic systems}} < \cdot \mu_{\text{C70,C60}} < \cdot \mu_{\text{CNT,QD}} \tag{6}$$

The observation of the increase of charge carrier mobility, high refractive index and high value of cubic nonlinearities predicts that the nonlinear optical and the dynamic feature of the nanostructured conjugated materials can be optimized via nanostructurization with good advantage. It should be noted that classical inorganic nonlinear volume media (including BSO, LiNbO$_3$, etc.) exhibit significantly lower nonlinearity, while bulk silicon based materials have nonlinear characteristics analogous to those of the organic thin film nanoobjects-doped materials under consideration.

Structure	Content of dopants, wt.%	Wavelength , nm	Energy density, Jcm^{-2}	Spatial frequency, mm^{-1}	Laser pulse duration, ns	Laser-induced change in the refractive index, (n	References
NPP	0	532	0.3	100	20	0.65x10^{-3}	[14]
NPP+C$_{60}$	1	532	0.3	100	20	1.65x10^{-3}	[14]
NPP+C$_{70}$	1	532	0.3	100	20	1.2x10^{-3}	[14]
PNP*	0	532	0.3	100	20	*	
PNP+C$_{60}$	1	532	0.3	100	20	0.8x10^{-3}	[14]
PI	0	532	0.6	90-100	10-20	10^{-4}-10^{-5}	[5]
PI+malachite green dye	0.2	532	0.5-0.6	90-100	10-20	2.87x10^{-4}	[5]
PI+graphene oxides	0.1	532	0.2	100	10	3.4x10^{-3}	present
PI+graphene oxides	0.2	532	0.28-0.3	100	10	3.65x10^{-3}	present
PI+shungites	0.1	532	0.6	150	10	3.1x10^{-3}	present
PI+shungites	0.2	532	0.063-0.1	150	10	5.3x10^{-3}	[9,15]
PI+C$_{60}$	0.2	532	0.5-0.6	90-100	10-20	4.2x10^{-3}	[5]
PI+C$_{70}$	0.2	532	0.6	90-100	10-20	4.68x10^{-3}	[5]
PI+C$_{70}$	0.5	532	0.6	90-100	10-20	4.87x10^{-3}	[5]
PI+C$_{70}$	0.1-0.5	1315	0.2–0.8	100	50	~10^{-3}	[14]

PI+quantum dots based on CdSe(ZnS)	0.003	532	0.2-0.3	100	10	2.0×10^{-3}	[10]
PI+CNTs	0.1	532	0.5-0.8	90-100	10-20	5.7×10^{-3}	[6]
PI+ CNTs	0.05	532	0.3	150	20	4.5×10^{-3}	[6,14]
PI+ CNTs	0.07	532	0.3	150	20	5.0×10^{-3}	[6,14]
PI+ CNTs	0.1	532	0.3	150	20	5.5×10^{-3}	[6,14]
PI + double-walled carbon nanotube powder	0.1	532	0.1	100	10	9.4×10^{-3}	[15]
PI + double-walled carbon nanotube powder	0.1	532	0.1	150	10	7.0×10^{-3}	[15]
PI+ mixture of CNT and nanofibers (type MIG)	0.1	532	0.3-0.6	90-100	10	11.7×10^{-3}	[15]
PI+ mixture of CNT and nanofibers (type MIG)	0.1	532	0.3-0.6	150	10	11.2×10^{-3}	[15]
Polymer-dispersed LC based on PI–C_{70} complex	0.2	532	0.1	90-100	10	1.2×10^{-3}	[10]
COANP	0	532	0.9	90-100	10-20	10^{-5}	[14]
COANP+ TCNQ**	0.1	676	2.2 Wcm^{-2}			2×10^{-5}	[16]
COANP+C_{60}	5	532	0.9	90-100	10-20	6.21×10^{-3}	[14]
COANP+C_{70}	0.5	532	0.6	100	10	5.1×10^{-3}	present
COANP+C_{70}	5	532	0.9	90-100	10-20	6.89×10^{-3}	[14]
Polymer-dispersed LC	0.5	532	30×10^{-3}	100	10	1.2×10^{-3}	present

based on COANP–C_{70} complex							
Polymer-dispersed LC based on COANP–C_{70} complex	5	532	17.5×10^{-3}	100	20	1.4×10^{-3}	[17]
Polymer-dispersed LC based on COANP–nanotubes	0.1	532	30×10^{-3}	100	10	2.8×10^{-3}	present
Polymer-dispersed LC based on COANP–nanotubes	0.5	532	18.0×10^{-3}	90-100	10-20	3.2×10^{-3}	[18]

Table 1. Laser-induced change in the refractive index in some organic structures doped with nanoobjects. * The diffraction efficiency has not detected for these systems at this energy density. ** Dye TCNQ - 7,7,8,8,-tetracyanoquinodimethane – has been used in the paper [16].

4. Conclusion

Thus, analysis of the obtained results leads to the following conclusions:

- Doping with nanoobjects significantly influences the photorefractive properties of nanobjects-doped organic matrices. An increase in the electron affinity (cf. shungite, fullerenes, QDs) and specific area (cf. QDs, CNTs, nanofilers) implies a dominant role of the intermolecular processes leading to an increase in the dipole moment, local polarizability (per unit volume) of medium, and mobility of charge carriers.

- A change in the distance between an intramolecular donor and intermolecular acceptor as a result of variation of the arrangement (rotation) of the introduced nanosensitizer leads to changes on the charge transfer pathway in the nanocomposite.

- The variations of the length, of the surface energy, of the angle of nanoobject orientation relative to the intramolecular donor fragment of matrix organics can significantly change the pathway of charge carrier transfer, which will lead to changes in the electric field gradient and dipole moment.

- Different values of nonlinear optical characteristics in systems with the same sensitizer type and concentration can be related to a competition between carrier drift and diffusion processes in a nanocomposite under the action of laser radiation.

- Special role of the dipole moment as a macroscopic parameter of a medium accounts for a relationship between the photorefraction and the photoconductivity characteristics.

- The photorefractive parameters change can be considered as the indicator of following dynamic and photoconductive characteristics change.

As the result of this discussion and investigation, new area of applications of the nanostructured materials can be found in the optoelectronics and laser optics, medicine, biology, telecommunications, display, microscopy technique, etc. Moreover, the nanostructured materials can be used for example, for development of 3D media with high density of recording information, as sensor in the gas storage and impurity testing, as photosensitive layer in the spatial light modulators, convertors, limiters, etc. devices.

Acknowledgements

The author would like to thank their Russian colleagues: Prof. N. M. Shmidt (Ioffe Physical-Technical Institute, St.-Petersburg, Russia), Prof. E.F.Sheka (University of Peoples' Friendship, Moscow, Russia), Dr.N.N.Rozhkova (Institute of Geology, Karelian Research Centre, RAS), Dr.A.I.Plekhanov (Institute of Automation and Electrometry SB RAS, Novosibirsk, Russia), Dr.V.I.Studeonov and Dr.P.Ya.Vasilyev (Vavilov State Optical Institute, St.-Petersburg, Russia), as well as foreign colleagues: Prof. Francois Kajzar (Université d'Angers, Angers, France), Prof. D.P. Uskokovic (Institute of Technical Sciences of the Serbian Academy of Sciences and Arts, Belgrade, Serbia), Prof. I.Kityk (Politechnica Czestochowska, Czestochowa, Poland), Dr. R.Ferritto (Nanoinnova Technologies SL, Madrid, Spain) for their help in discussion and study at different their steps. The presented results are correlated with the work supported by Russian Foundation for Basic Researches (grant 10-03-00916, 2010-2012).

Author details

Natalia V. Kamanina[1*]

Address all correspondence to: nvkamanina@mail.ru

1 Vavilov State Optical Institute, 12, Birzhevaya Line, St. Petersburg, , Russia

References

[1] Couris, S., Koudoumas, E., Ruth, A. A., & Leach, S. (1995). Concentration and wavelength dependence of the effective third-order susceptibility and optical limiting of C_{60} in toluene solution. *J. Phys. B: At. Mol. Opt. Phys.*, 8, 4537-4554.

[2] Robertson, J. (2004). Realistic applications of CNTs. *.Mater. Today*, 7, 46-52.

[3] Lee, Wei, & Chen, Hsu-Chih. (2003). Diffraction efficiency of a holographic grating in a liquid-crystal cell composed of asymmetrically patterned electrodes. *Nanotechnology*, 14, 987-990.

[4] Buchnev, Oleksandr, Dyadyusha, Andriy, Kaczmarek, Malgosia, Reshetnyak, Victor, & Reznikov, Yuriy. (2007). Enhanced two-beam coupling in colloids of ferroelectric nanoparticles in liquid crystals. J. Opt. Soc. Am. B/ , 24(7), 1512-1516.

[5] Kamanina, N. V., Emandi, A., Kajzar, F., & Attias, Andre'-Jean. (2008). Laser-Induced Change in the Refractive Index in the Systems Based on Nanostructured Polyimide: Comparative Study with Other Photosensitive Structures. *Mol. Cryst. Liq. Cryst.*, 486, 1-11.

[6] Kamanina, N. V., Serov, S. V., Savinov, V. P., & Uskoković, D. P. (2010). Photorefractive and photoconductive features of the nanostructured materials. *International Journal of Modern Physics B (IJMPB)*, 24(6-7), 695-702.

[7] Kamanina, N. V., & Vasilenko, N. A. (1997). Influence of operating conditions and of interface properties on dynamic characteristics of liquid-crystal spatial light modulators. *Opt. Quantum Electron.*, 29, 1-9.

[8] Kamanina, N. V. (2005). Fullerene-dispersed liquid crystal structure: dynamic characteristics and self-organization processes. *Physics-Uspekhi*, 48, 419-427.

[9] Kamanina, N. V., Serov, S. V., Shurpo, N. A., & Rozhkova, N. N. (2011). Photoinduced Changes in Refractive Index of Nanostructured Shungite-Containing Polyimide Systems. *Technical Physics Letters*, 37(10), 10(10), 949-951.

[10] Kamanina, N. V., Shurpo, N. A., Likhomanova, S. V., Serov, S. V., Ya, P., Vasilyev, V. G., Pogareva, V. I., Studenov, D. P., & Uskokovic, . (2011). Influence of the Nanostructures on the Surface and Bulk Physical Properties of Materials. *ACTA PHYSICA POLONICA A*, 119(2), 2(2), 256-259.

[11] Kamanina, N. V., & Plekhanov, A. I. (2002). Mechanisms of optical limiting in fullerene-doped conjugated organic structures demonstrated with polyimide and COANP molecules. *Optics and Spectroscopy*, 93(3), 408-415.

[12] Cherkasov, Y. A., Kamanina, N. V., Alexandrova, E. L., Berendyaev, V. I., Vasilenko, N. A., & Kotov, B. V. (1998). Polyimides: New properties of xerographic, thermoplastic, and liquid-crystal structures. (SPIE International Symposium on Optical Science, Engineering and Instrumentation, San Diego, CA, USA, 1998) Proceed. of SPIE , 3471, 254-260.

[13] Akhmanov, S. A., & Nikitin, S. F. (1998). Physical Optics. (Izdat. Mos. Gos. Univ., Moscow, 1998) [in Russian].

[14] Kamanina, N. V., Vasilyev Ya, P., Serov, S. V., Savinov, V. P., Bogdanov. Yu, K., & Uskokovic, D. P. (2010). Nanostructured Materials for Optoelectronic Applications. *Acta Physica Polonica A*, 117(5), 786-790.

[15] Kamanina, N. V., Serov, S. V., Shurpo, N. A., Likhomanova, S. V., Timonin, D. N., Kuzhakov, P. V., Rozhkova, N. N., Kityk, I. V., Plucinski, K. J., & Uskokovic, D. P. (2012). Polyimide-fullerene nanostructured materials for nonlinear optics and solar energy applications. J Mater Sci: Mater Electron DOI: 10.1007/s10854-012-0625-9, published on-line 26 January 2012.

[16] Ch Bosshard. Bosshard., K., Sutter, P., & Chapuis. Günter, G. (1989). Linear- and non-linear-optical properties of 2 -cyclooctylamino-5-nitropyridine. *J. Opt. Soc. Am.*, B6, 721-725.

[17] Kamanina, N. V. (2002). Optical investigations of a C_{70}-doped 2-cyclooctylamino-5-nitropyridine-liquid crystal system. *Journal of Optics A: Pure and Applied Optics,*, 4(4), 4(4), 571-574.

[18] Kamanina, N. V., & Uskokovic, D. P. (2008). Refractive Index of Organic Systems Doped with Nano-Objects. *Materials and Manufacturing Processes*, 23, 552-556.

[19] Chemla, D. S., & Zyss, J. (1987). Nonlinear Optical Properties of Organic Molecules and Crystals . (Orlando Academic Press, 1987), Translated into Russian, Moscow: Mir, 1989., 2

[20] Gutman, F., & Lyons, L. E. (1967). Organic Semiconductors. Wiley, New York.

[21] Mylnikov, V. S. (1994). in Advances in Polymer Science. Photoconducting Polymers/ Metal-Containing Polymers Springer-Verlag Berlin , 115, 3-88.

[22] Wang, Y., & Suna, A. (1997). Fullerenes in Photoconductive Polymers. Charge Generation and Charge Transport. *J. Phys. Chem.*, B 101, 5627-5638.

Chapter 6

Carbon Nanotube-Enzyme Biohybrids in a Green Hydrogen Economy

Anne De Poulpiquet, Alexandre Ciaccafava,
Saïda Benomar, Marie-Thérèse Giudici-Orticoni and
Elisabeth Lojou

Additional information is available at the end of the chapter

1. Introduction

Alternative energy pathways to replace depleting oil reserves and to limit the effects of global warming by reducing the atmospheric emissions of carbon dioxide are nowadays required. Dihydrogen appears as an attractive candidate because it represents the highest energy output relative to the molecular weight (120 MJ kg^{-1} against 50 MJ kg^{-1} for natural gas), and because its combustion delivers only water and heat. Whereas the main renewable sources of energy available in nature (solar, wind, geothermal…) need to be transformed, dihydrogen is able to transport and store energy. Dihydrogen can be produced from renewable energies, indirectly from photosynthesis *via* biomass transformation, or directly by bacteria. It can be converted into electricity using fuel cell technology. From all these properties and because it does not compete with food and water resources, dihydrogen has been defined as third generation biofuel. It thus emerges as a new fully friendly environmental energy vector. The use of dihydrogen as an energy carrier is not a new idea. Let us simply remember that Jules Verne, a famous French visionary novelist, wrote early in 1874: "I believe that O_2 and H_2 will be in the future our energy and heat sources" [1]. His prediction simply relied on the discovery a few years before of the fuel cell concept by C. Schönbein, then W. Groove, who demonstrated that when stopping water electrolysis, a current flow occurred in the reverse way [2]. However in order to implement the dihydrogen economy and replace fossil fuels, there are significant technical challenges that need to be overcome in each of the following domains:

1. dihydrogen production and generation,

2. dihydrogen storage and transportation,

3. dihydrogen conversion to electrical energy.

As opposed to widespread opinions, natural dihydrogen sources exist alone on the earth's surface. Local and continuous emanations of dihydrogen can be observed in cratonic zones, ophiolitic rocks or oceanic ridges [3]. Dihydrogen is effectively produced in the upper mantle of the earth through natural oxidation of iron (II)-rich minerals, like ferromagnesians, by water of the hydrosphere. The ferrous iron is oxidized in ferric iron and water is concurrently reduced in dihydrogen, as given by following equation: $2Fe^{2+}$ (mineral) + 2 H^+ (water) → 2 Fe^{3+} (mineral) + H_2. The same reaction can occur with other ions like Mn^{2+}. Exploitation of these sources remains however difficult so far as dihydrogen does not accumulate on the earth subsurface, especially for two reasons. First because as a powerful energy source dihydrogen is quickly consumed (biologically or abiotically), and second because as the lightest and most mobile gas it is not much retained by Earth's attraction and escapes in the atmosphere.

Combined with water and hydrocarbons dihydrogen is nevertheless the most abundant element on earth. Green means to ecologically convert H containers into dihydrogen still remain however challenging. The energetic volume density of dihydrogen is low (10.8 MJ m^{-3} against 40 MJ m^{-3} for natural gas) so that storage and transportation appear as bottlenecks for large scale development in transportation for example. Conversion of dihydrogen to electricity in fuel cells presents high electrical efficiency (more than 50% against less than 30% for gas engines), but requires the use of catalysts both for H_2 oxidation and O_2 reduction. These are mainly based on platinum catalysts, which are highly expensive, weakly available on earth, and non biodegradable. Extensive researches thus aim to decrease the amount of platinum catalysts in fuel cells. Following the discovery of carbon nanotubes (CNTs) [4, 5], their large scale availability opened a new avenue in these three domains. Due to their intrinsic properties, such as high stability, high electrical and thermal conductivities [6] and high developed surface areas, carbon nanotubes constitute attractive materials, able to enhance the credibility of an hydrogen economy.

Besides, platinum catalysts are inhibited by very low amount of CO and S (0.1% of CO is sufficient to decrease one hundred fold the catalytic activity of Pt in ten minutes!), thus requiring strong steps of H_2 purification [7]. They are not specific to either O_2 or H_2 catalysis, thus requiring the use of a membrane to separate the anodic and cathodic compartments. Nafion® perfluoronated membrane is currently the only really performing polymer [8], increasing its cost. Replacement of platinum-based catalysts is thus highly needed. In that way, a new concept appeared less than five years ago, when looking at the pathways microorganisms use for the production of ATP, their own energetic source [9, 10]. As an example, the hyperthermophilic, microaerophilic bacterium *Aquifex aeolicus*, couples H_2 oxidation to O_2 reduction *via* a membrane quinone pool (Figure 1). The redox coupling generates a proton gradient through the cell membrane for ATP synthesis. Clearly, this pathway can be considered as an *"in vivo* biofuel cell". The question rises if we could take benefit of bacterial energetic pathways for our own energetic needs. The idea thus emerged that microorganisms or enzymes could be used instead of chemical catalysts for the development of efficient electricity producing devices. These innovative batteries called biofuel cells rely on enzymes highly specific for various fuels and oxidants [11]. A mandatory condition is that these enzymes have to be immobilized onto electrodes. One of the most common biofuel cell uses glucose oxidase and laccase, two enzymes specific for glucose oxidation and oxygen reduction, respectively. A few years ago, a new concept of biofuel cells appeared based on enzymes specific of dihydrogen oxidation. This biohydrogen economy relies on the opportunity to use low-cost materials for efficient conversion of solar energy to dihydrogen and of dihydrogen to electricity. Many microorganisms biosynthesize hydrogenase, the metalloenzyme that catalyzes the dihydrogen conversion. At least two modes of application of dihydrogen-metabolizing protein catalysts are nowadays considered within dihydrogen as a future energy carrier. Hydrogenases may be used as catalysts in dihydrogen production by coupling oxygenic photosynthesis to biological dihydrogen production [12]. Hydrogenases can also be used directly as anode catalysts in biofuel cells instead of chemical catalysts [13]. The improved knowledge of hydrogenase structure and of catalytic mechanisms allows nowadays to design the development of biofuel cells functioning as Proton Exchange Membrane (PEM) fuel cells.

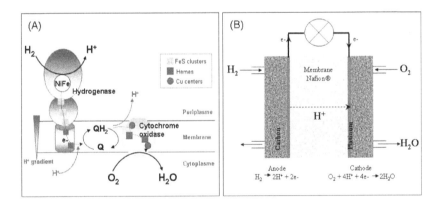

Figure 1. A) Energetic metabolism of the bacterium *Aquifex aeolicus*: H_2 oxidation in the periplasm is coupled to O_2 reduction in the cytoplasm via a membrane quinone pool to generate a trans membrane proton gradient for ATP synthesis; (B) General view of a chemical PEM fuel cell.

For all these innovative concepts, one of the key points is the increase in power density, thus in the current density furnished by a redox couple displaying a large as possible potential difference. Apart from the improvement in enzyme stability, the increase in the current densities supposes an optimization of both the interfacial electron transfer rate and the amount of connected enzymes at the electrode. Carbon nanotubes which develop large surface areas and can be functionalized constitute an attractive platform for such enzyme immobilization. CNTs are described as graphene sheets rolled into tubes. They exist under various structural configurations (single-walled (SWCNTs), multi-walled (MWCNTs)) differing in electrical properties, thus tuning the platform properties for enzyme immobilization. The end of the tubes is capped by a fullerene-type hemisphere that yields selective functionalization of the CNTs [14].

With the objective of dihydrogen as a future green energy vector, this review focuses on the last developments in the fuel -and more especially biofuel- cell field thanks to the advantageous use of carbon nanotubes. In a first part, carbon nanotubes for H_2 storage enhancement are discussed. Then fuel cells in which carbon nanotubes help to decrease the amount of high cost noble metal catalysts are described. Green H_2 economy is then emphasized considering the key role of hydrogenase, the enzyme responsible for dihydrogen conversion. This requires the functional immobilization of the biocatalysts onto electrodes. The use of carbon nanotubes in this immobilization step is underlined, including the modes of carbon nanotube functionalization and enzyme or microbes grafting. Then the advantages of developing biofuel cells in which chemical catalysts are replaced by enzymes or microbes are described. A short review of the sugar/O_2 biofuel cells, the most widely investigated biofuel cell, is given with a particular attention on the devices based on carbon nanotube-modified bioelectrodes. The last developments based on carbon nanotube networks for hydrogenase immobilization, or mimicking synthetic complex immobilization, in view of efficient dihydrogen catalytic oxidation are finally described in order to allow the design of a future H_2/O_2 biofuel cell.

2. Carbon nanotubes: an attractive carbon material

The discovery of carbon nanotubes (CNTs) has induced breakthroughs in many scientific domains, including H_2 economy, biosensors, bioelectrochemistry…This is due to their remarkable properties, such as good electronic, mechanical and thermal properties. Their nanometric size compares with that of proteins and enzymes, offering the possibility of electrical connection. Their large developed surface area allows the development of devices in smaller volumes. SWCNTs are sp2 hybridized carbon in a hexagonal honeycomb structure that is rolled into hollow tube morphology [15]. MWCNTs are multiple concentric tubes encircling each other [5]. Depending on the chirality, CNTs can be metallic or semiconducting. The distinction between metallic and semiconducting is very important for application, but the physical separation of allotropes is one of the most difficult challenges to overcome. In MWCNTs, a single metallic layer results in the entire nanotubes metallic behavior. Most often mixtures of these two forms are present in CNTs preparation. More information on the physical and electronic structures can be found in many published reviews [16]. CNTs are produced by various methods such as arc discharge, laser ablation, and chemical vapor deposition (CVD). Commercially CNTs are generally produced by CVD during the pyrolysis of hydrocarbon gases at high temperature. The control of synthesis parameters (reagent gas, T °, metal catalysts) allows for the control of CNT properties. Metal impurities may remain in the CNTs sample, thus requiring purification steps. CNTs may be treated to functionalize the surface.

3. Carbon nanotubes for safe and efficient H_2 storage

The use of H_2 in fuel cells to generate electricity has been proved early in the middle of the nineteenth century. Surprisingly this discovery by C. Schönbein in 1839 of current generation by use of H_2 and O_2 in sulphuric acid was applied by NASA only late in 1960. Despite intensive studies over the last two decades, fuel cells still suffer from high cost and low durability. The first difficulty responsible for this slow large scale development lies on dihydrogen storage and transportation, both regarded as bottlenecks considering dihydrogen specific volumic density as a gas. For convenience the gas must be intensely pressurized to several hundred atmospheres and stored in a pressure vessel. The ways to store dihydrogen with minimum hazard are under liquid state under cryogenic temperatures (at a temperature of -253 °C), or more efficiently in a solid state. Storage of dihydrogen in hydride form uses an alloy that can absorb and hold large amounts of dihydrogen by bonding with hydrogen and forming hydrides. A dihydrogen storage alloy is capable of absorbing and releasing dihydrogen without compromising its own structure, according to the reaction: M + $H_2 \leftrightarrow MH_2$, where M represents the metal and H, hydrogen. Qualities that make these alloys useful include their ability to absorb and release large amounts of dihydrogen gas many times without deteriorating, and their selectivity toward dihydrogen only. In addition, their absorption and release rates can be controlled by adjusting temperature or pressure. The dihydrogen storage alloys in common use occurs in four different forms: AB_5 (e.g., $LaNi_5$), AB (e.g., FeTi), A_2B (e.g., Mg_2Ni) and AB_2 (e.g., ZrV_2). Metal hydrides, such as MgH_2, Mg_2NiH_4 or $LiBH_4$, constitute secure reserves of dihydrogen [17-19]. Dihydrogen is released from MH_2 upon increase in temperature and/or decrease in pressure.

Material	H_2 gas, 200 bar	H_2 liquid, -253 C	MgH_2	Mg_2NiH_4	$FeTiH_2$	$LaNi_5H_6$
H-atom per cm^3 ($\times 10^{22}$)	0.99	4.2	6.5	5.9	6.0	5.5

Table 1. H density as a function of storage method.

Much progress has been made during the last years in that domain, including the highlight of the advantages offered by using CNTs. An efficient approach appears to be the formulation of new carbon/transition metal catalyst composites of specific composition and molecular structure, which can greatly stimulate and improve the chemical reactions involving dihydrogen relocation in alkali-metal aluminium materials. Absorption kinetics and dihydrogen storage capacity were shown to be enhanced by mixing MH_2 with SWCNTs as a result of an increase in interfacial area, decrease in MH_2 particle agglomeration and nanoplatform for efficient H_2 diffusion [20, 21]. The hydriding and dehydriding kinetics of SWCNT/catalyzed sodium aluminium composite were found to be much better than those of the material ground without carbon additives. Temperature of H_2 desorption was lowered [22]. The presence of carbon creates new dihydrogen transition sites and the high dihydrogen diffusivity of the nanotubes facilitates hydrogen atom transition. Faster thermal energy transfer through the nanotubes may also help reduce hydriding and dehydriding times.

Dihydrogen can be stored through physisorption on CNTs, based on Van der Waals interaction. Based on the surface area of a single graphene sheet, the maximum value for the storage of dihydrogen capacity is around 3 wt%. Dihydrogen can also be stored through chemisorption in CNTs matrix. If the π-bonding between carbon atoms were fully utilized, every carbon atom could be a site for chemisorption of one hydrogen atom. Dillon et al. first reported in 1997 dihydrogen storage in SWCNT networks [23]. Both SWCNTs and MWCNTs store dihydrogen in microscopic pores on the tubes [24, 25]. Similar to metal hydrides in their mechanism for storing and releasing dihydrogen, the carbon nanotubes hold the potential to store a significant volume of dihydrogen. The storage capacity is dependent on many parameters of the CNTs, including their structure, structure defects, pretreatment, purification, geometry (surface area, tube diameter, length), arrangement of tubes in bundles, storage pressure, temperature,…Dihydrogen uptake varies linearly with tube diameter, because the uptake is proportional to the surface area, *i.e.* the number of carbon atoms. The adsorption sites exist inside and outside the tube, between tubes in bundles, between the shells in MWCNTs. For dihydrogen storage into the tube dihydrogen must pass through the CNT wall or the tube must be opened. Hydrogen forms stable C-H bonds on SWCNT surface at room temperature that can dissociate above 200°C. According to SWCNT diameter 100% hydrogenation can be obtained, thus more than 7 wt % dihydrogen storage capacity, which is above the target fixed by the US Department of Energy's Office of Energy Efficiency and Renewable Energy [26].

4. Carbon nanotubes for a decrease in the amount of noble metal catalysts in fuel cells

Among the different types of fuel cells, PEM fuel cell operates at low temperatures around 100°C. For small portable application requiring less than 10 kW, they are more suitable than higher powering solid oxide fuel cells (functioning at 700°C) due to the possible use of usual materials for electronic connectors (mainly based on carbon) and membrane. However the necessary use of platinum-based catalysts on electronic connectors to accelerate the rate of dihydrogen oxidation and oxygen reduction is a real brake towards the fuel cell development. Platinum is scarce enough on earth to be a limiting factor in case of large scale development of fuel cells. Consequently platinum currently accounts for 25% in the total cost of a fuel cell. Over the past five years, the price of platinum has ranged from just below $800 to more than $2,200 an ounce. Carbon black particles offer a high surface area support, able to decrease the amount of platinum particles. But they suffer from mass transfer limitations and strong carbon corrosion.

Among the low-cost alternatives to platinum, carbon appears to be the most promising. Due to their nano-structure and unique chemical and physical properties, CNTs have appeared as ideal supporting materials to improve both catalytic activity and electrode stability. The enhancement of fuel cell performances by using CNT/Pt or Pt-alloy catalysts may arise from:

i.   higher dispersion of Pt nanoparticles,

ii.   increased electron transfer rates,

iii.   porous structure of CNT layers.

Various CNT-Pt composites were used to reduce the platinum amount while preserving high catalytic activity in PEM fuel cells. Platinum nanodots sputter-deposited on a CNT-grown carbon paper [27], or deposited on functionalized MWCNTs [28] exhibited great improvement in cell performance compared to platinum on carbon black. This was primarily attributed to high porosity and high surface area developed by the CNT layer. Compared to a commercial Pt/carbon black catalyst, Pt/SWCNT films cast on a rotating disk electrode was shown to exhibit a lower onset potential and a higher electron-transfer rate constant for oxygen reduction. Improved stability of the SWCNT support was also confirmed from the minimal change in the oxygen reduction current during repeated cycling over a period of 36 h [29]. Platinum particles deposited on MWCNT encapsulated in micellar surfactant were also explored as efficient catalysts for fuel cells [30, 31]. An in situ synthetic method was reported for preparing and decorating metal nanoparticles at sidewalls of sodium dodecyl sulfate micelle functionalized SWCNTs/MWCNTs. Accelerated durability evaluation was carried out by conducting 1500 potential cycles between 0.1 and 1.2 V at 80°C. These nanocomposites were demonstrated to yield a high fuel cell performance with enhanced durability. The membrane electrode assembly with Pt/MWCNTs showed superior performance stability with a power density degradation of only 30% compared to commercial Pt/C (70%) after potential cycles. Identically electrocatalytically active platinum nanoparticles on CNTs with enhanced nucleation and stability have been demonstrated through introduction of electron-conducting polyaniline (PANI) [32]. A bridge between the Pt nanoparticles and MWCNTs walls was demonstrated with the presence of platinum nitride bonding and π-π bonding. The synthesized PANI was found to wrap around the CNT as a result of π-π bonding, and highly dispersed Pt nanoparticles were loaded onto the CNT with narrowly distributed particle sizes ranging from 2.0 to 4.0 nm. The Pt-PANI/CNT catalysts were electroactive and exhibited excellent electrochemical stability, therefore constitute promising potential applications in proton exchange membrane fuel cells. Strong evidence thus emerges that CNTs/Pt composites are efficient as catalysts for fuel cells. Although platinum content has been dramatically decreased, industrials consider that further optimization is mandatory for a large scale fuel cell production. In addition Nafion® membrane between the cathodic and anodic compartment delays the large scale application of fuel cells, due to cost and problem of mass transfer. Breakthrough research towards these two bottlenecks could surely enforce a hydrogen economy.

5. Towards a green H₂ economy: carbon nanotubes for enzyme and microbe immobilization

Replacement of chemical catalysts is thus nowadays highly needed in view of the development of a green energy economy. Microorganisms contain many biocatalysts, namely enzymes, which are highly efficient and specific towards various substrate conversions. Given they are produced in large enough quantities, these enzymes could be used as catalysts in biotechnological devices. A mandatory condition to develop heterogeneous catalysis is to succeed in the functional immobilization and in the stabilization of the enzymes on solid supports. The redox active site of enzymes is indeed buried inside the protein moiety so that the enzymatic property can be maintained under environmental stresses. Specific channels are often involved to allow the substrate to reach the active site. Complex but highly organized electron transfer chains occur for energetic metabolism. Electron transfer between two physiological partners associated with transformation of the substrate involves specific recognition site. The game for a bioelectrochemist that aims to get the highest electron transfer rate for heterogeneous catalysis is to reproduce at the electrode interface the physiological electron transfer recognition process. Given the usual size of an enzyme (5-10 nm), electron transfer cannot occur *via* electron tunneling from the active site to the surface of the enzyme. In some enzymes, electron relays, one being located at the protein surface, act as a conductive line for electron shuttling. If the electrode interface is built so that it fits the surface electron relay environment, one can expect to favor a direct electrical connection of the enzyme onto the electrode. In case of direct electron transfer failure, an artificial redox mediator that acts as a fast redox system and shuttles electrons between the enzyme and the electrode can be used (Figure 2) [13, 33].

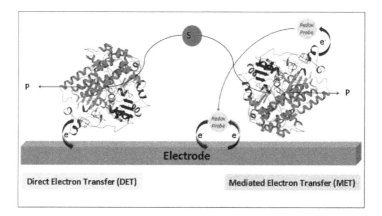

Figure 2. Interfacial electron transfer between an enzyme and an electrode can be achieved by direct (left) or mediated (right) electron transfer process.

Direct electron transfer process is preferred to mediated one, because it is not limited by the affinity between the enzyme and the redox mediator, and because it avoids the co-immobilization of enzyme and mediator. It is furthermore expected to yield the highest power density because enzymes, as biocatalysts, transform their substrate into products with very low overvoltages. However it requires the knowledge of the protein structure and the construction of a tuned electrochemical interface that fits the electron transfer site.

There are many strategies for efficient enzyme immobilization onto electrochemical interfaces, including simple physical adsorption, covalent attachment, cross-linking or entrapment in polymers. The objectives are to optimize the immobilization procedure so that the efficiency of the enzyme and its stability are preserved. Moreover, due to the size of enzymes compared to chemical catalysts, large surface area interfaces baring many anchorage sites are required to obtain high catalytic currents.

To reach these goals, 3D structures are preferred, and CNT-based electrodes are very popular, both SWCNTs and MWCNTs. CNTs can be directly grown onto electrode surface, or adsorbed on it, or imbedded in polymer coating. In most cases, higher activity was reported for enzymes physically adsorbed onto CNTs [34]. Hydrophobic interactions between the enzyme and the CNT walls and π-π interactions between side walls of CNTs and aromatic rings of the enzyme are thought to be the driving force for direct adsorption of enzymes on CNTs [35]. Electrostatic interaction between the defect sites of CNTs and protonated amino residues of the enzyme plays also a role in the adsorption process [35]. CNTs are quite easily functionalized, allowing covalent, thus stable specific attachment of enzymes. The oxidation in strong acidic solutions at high temperature was demonstrated to remove the end caps and shorten the lengths of the CNTs. The length of the CNTs was shown to be a function of the oxidation duration [36]. Acid treatment also adds oxide groups, primarily carboxylic acids, to the tube ends and defect sites [37]. The control of reactants and/or oxidation conditions may control the locations and density of the functional groups on the CNTs, which can be used to control the location and density of the attached enzymes [37]. Covalent immobilization is induced by carbodiimide reaction between the free amine groups on the enzyme surface and carboxylic groups generated by side wall oxidation of CNTs.

Further chemical reactions can be performed at the oxide groups generated on the oxidized CNTs to functionalize with groups such as amides, thiols, etc...From an electrochemical point of view, the side walls of CNTs were suggested to behave as basal plane of pyrolytic graphite, while their open ends resemble the edge planes [38, 39]. But recent work has demonstrated that the side wall may be responsible for electrochemical activity [40]. It has been furthermore suggested that the uncovered surface of CNTs promotes the accessibility of the substrate to the enzyme [41]. It is also interesting to note that the open spaces between CNTs are accessible to large species such as entire bacteria [42], opening the way for the development of fuel cells using whole microorganisms instead of purified enzymes. The cost and complexity of CNT manufacturing seem to be still clogging issues in that field.

Abundant literature exists on the ways CNTs are architectured for efficient enzyme immobilization, including those specific for development of enzymatic fuel cells. Enzymes and proteins as various as glucose oxidase and dehydrogenase, tyrosinase, laccase and bilirubin oxidase, peroxidase, haemoglobin and myoglobin, *i.e.* flavin, copper or heme containing active sites, have been studied. Whereas direct electron transfer between protein or enzyme and an electrochemical interface has been for long time supposed to be restricted to small proteins (<15kDa) possessing active sites exposed to the surface (it is the case for many cytochromes as example [43]), the use of CNT-modified electrodes has greatly enhanced the number and kinds of enzymes able to be directly connected to an electrode. Enzymes as large as one hundred kDa, with many cofactors are now considered for direct electron transfer. Consequently, recent works during the last years focus and report on direct communication between enzymes and electrode interface through CNT network. The induced porosity of the film depends on the type of CNTs used. But generally the nanometric size of the CNTs compared to the size of enzymes favors a direct electronic connection of the enzyme whatever its orientation [44]. The physical properties of CNTs, including high electrical conductivity, explain why CNT layers can be built up on electrodes most often yielding high rate direct electron transfer for enzymatic product transformation. Many researches report on the increase in electroactive surface area by use of CNT coatings that contribute to an increase in the direct electron transfer process [45-52]. CNTs are usually deposited on electrodes as thick films. Alternatively, layer-by-layer (LBL) process induces a quite stable protein film with nice electrocatalytic properties [53-55]. LBL is based on electrostatic interaction between oppositely charged monolayers in an alternating assembling. Although CNTs greatly amplify the current response, layer-by-layer architecture suffers from weak stability of the build-up and decrease in electron transfer for the upper layers. Besides vertically aligned CNTs were suggested to act as molecular wires that ensure the electrical communication between enzyme and electrode [56-58]. The carboxylic functions induced by acidic treatment of CNTs can be used for further chemical modifications. Amine- [59-61], thionin- [62, 63], diazonium salts [64, 65], pyrene [66, 67] (Figure 3) or other π-π stacking interactions [68] were used to functionalize CNTs. These modifications were demonstrated to be efficient platforms for enzyme immobilization.

Mixing CNTs with surfactant [69-71] was claimed to assist in the dispersion of CNTs while avoiding oxidative functionalization which may disrupt their π-network. Polymer modified CNTs [72, 73] and sol-gel-CNT nanocomposite films [74] were proved to behave as friendly platforms for enzyme encapsulation.

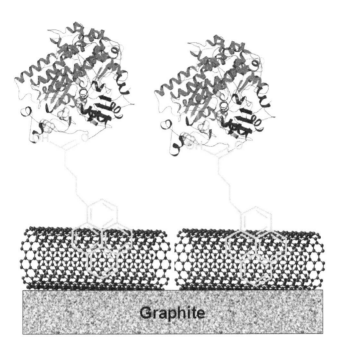

Figure 3. Schematic drawing of the build-up of enzyme on SWCNTs *via* π-π interactions.

Many enzymes however cannot be electrically connected to the electrode interface and require redox mediator to electrochemically follow substrate conversion. In that case, electrode kinetics is mainly dependant on mediator kinetics, so that the choice of the redox mediator mainly impedes the power density. Another issue is that the mediator can be co-immobilized with the enzyme at the electrode, while still being capable of efficient interaction with the enzyme. CNTs have also been used for building networks enabling co-immobilization of enzymes and redox mediators. In that way, one of the most popular redox entities is osmium polymer which forms hydrogels with enzymes allowing both charge transfer reaction between enzyme and mediators and diffusion of substrate and product [75]. Composite CNT/osmium films were used To immobilize bacteria [76], or enzymes [77]. By optimizing the CNT and polymer amounts, enhanced current responses were obtained linked to a promotion of the electron transfer within the composite. Various phenothiazine derivatives were also used to form nanohybrids with CNTs acting as efficient redox mediator platforms [78-80]. Phenothiazine derivatives strongly adsorb onto CNTs leading to great enhancement of redox dye loading onto the electrode, but also to improved electrochemical sensing devices. Another strategy involves the use of a redox polymer as redox mediator platform. Electropolymerization of the redox conducting polymer onto CNTs enhances the amount of redox units and the electrical conductivity of the coating [81]. An interesting construction has also been obtained by immobilization of physiological cofactor onto CNT layers *via* π-π interactions, then immobilization of the enzyme [82]. The covalent coupling between the enzyme and its natural cofactor which was immobilized onto CNTs was proved to be efficient towards mediated substrate catalysis. This overview of multiple architectures involving enzymes and CNTs highlights the deep efforts engaged in the last years for efficient biocatalyst immobilization that open avenues towards biotechnological devices.

6. Carbon nanotubes for biological production of dihydrogen

Apart from replacement of noble metal catalysts in fuel cells, a new green technology for production of dihydrogen is required. It currently relies on steam reforming of hydrocarbons under high temperature and pressure conditions, which starts from fossil fuels, thus producing greenhouse gases. Dihydrogen production *via* water electrolysis appears as a renewable solution given that the energy input comes from a renewable source, ideally solar energy. Many bacteria gain energy by the oxidation of dihydrogen assisted by a number of complex mechanisms. Various species evolve H_2 under anaerobic conditions. This is also a human being process since bacteria in our digestive tract produce H_2, though not detectable because immediately recycled by other bacteria. Photosynthetic organisms such as microalgae and cyanobacteria are very efficient in water splitting [83]. They possess photosensitizers for photon capture and charge separation, and enzymes for water oxidation to oxygen and water reduction to dihydrogen. This chemical activity relies on the expression of very efficient enzymes, called hydrogenases [84], which catalyze with high turn-over (one molecule of hydrogenase produces up to 9000 molecules of H_2 per second at neutral pH and 37°C) and low overvoltage the conversion of protons into dihydrogen and the oxidation of dihydrogen. The sequences of 450 hydrogenases are now available. Hydrogenases differ in size, structure, electrons donors. They also differ by their position in the cell (soluble in the periplasm, membrane-bound), and by their activity preferentially towards H_2 oxidation or protons reduction. Hydrogenase active site is composed of non noble metals such as iron and nickel, unlike platinum catalyst necessary for the chemical electrolysis of water. Three distinct classes can be split which differ from the type of metal content in the active site: [NiFe], [FeFe] and [Fe] hydrogenases. [NiFe] and [FeFe] hydrogenases possess dinuclear active centers which are connected through thiolate bridges. [NiFe] hydrogenase (Figure 4) is the most usual hydrogenase in microorganisms. It is composed of two subunits. The larger subunit harbors the [NiFe] active site. The small subunit contains FeS clusters. Electrons are transferred to the active site along these FeS clusters distant less than 10 Å that act as a conductive line. [FeFe] hydrogenases are monomeric. In addition to the active site they contain additional domains which accommodate FeS clusters.

In order to use these biocatalysts for green dihydrogen production, two main research domains are currently concerned: the understanding of the catalytic mechanisms of H_2 production, and the optimization of enzyme immobilization. Adsorption onto graphite electrodes [85, 86] was largely used to study the mechanisms by which hydrogenases produce H_2.

Grafting of hydrogenase onto gold electrode modified by thiolated Self-Assembled-Mono-layer [87] allowed efficient proton reduction into dihydrogen in aqueous buffer solutions. Hydrogenase is also considered as a promising biocatalyst for photobiological production of dihydrogen when coupled to a photocatalyst [88]. Hybrid complexes of hydrogenases with TiO_2 nanoparticles [89, 90] were studied for H_2 production. The optimized system was shown to produce H_2 at a turnover frequency of approximately 50 (mol H_2) s^{-1} (mol total hy-drogenase)$^{-1}$ at pH 7 and 25 °C, even under the typical solar irradiation of a northern Euro-pean sky. Cd-based nanorods [91, 92] were recently studied. The CdS nanorod/hydrogenase complexes photocatalyzed reduction of protons to H_2 at a hydrogenase turnover frequency of 380-900 s^{-1} and photon conversion efficiencies of up to 20% under illumination at 405 nm. Cd-based complexes allowed photoproduction of dihydrogen for a couple of hours, but still suffer from quick inhibition of hydrogenase.

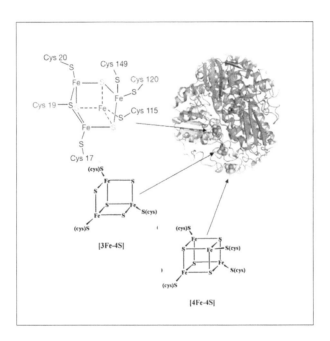

Figure 4. Structure of an oxygen-tolerant [NiFe] hydrogenase.

Although a very attractive way, little work has been done towards enhancement of green hydrogen production using CNTs. Three studies from the same group reported however catalytically active hydrogenase-SWCNT biohybrids [93, 94]. Surfactant-suspended SWCNTs were shown to spontaneously self-assemble with hydrogenase. Photolumines-cence excitation and Raman spectroscopy showed that SWCNTs act as molecular wires to make electrical contact with at least one of the FeS electron relay. Hydrogenase was demonstrated to be strongly attached to the SWCNTs and to mediate electron injection into nanotubes. The displacement of the surfactant by hydrogenase to gain access to the SWCNTs was strongly suggested by photoluminescence studies. Furthermore, Raman studies of charge transfer complexes between hydrogenase and either metallic (m) or semiconducting (s) SWCNTs revealed a difference in oxygen deactivation of hydrogenase according to the SWCNT species. m-SWCNTs most probably interact with hydrogenase to produce a more oxygen-tolerant species. The study further suggested that purified m-SWCNTs or s-SWCNTs, rather than mixed preparation, would be more suitable for hydrogenase-SWCNTs biohybrids. The formation of these catalytically active biohybrids in addition with the intrinsic properties developed by CNT networks on electrodes certainly accounts for the improved dihydrogen production observed in the following studies. Kihara *et al.* immobilized hydrogenase on a SWCNT-forest with a unique dense structure of vertically aligned millimetre-scale height SWCNTs [95]. Hydrogenase was demonstrated to spontaneously assemble between adjacent nanotubes. The maximum rate of dihydrogen production was reported to be 720 nmol/min/(mg hydrogenase) and the electron transfer efficiency was estimated to be 32%. It is two thousand fold higher than reported before using the same hydrogenase on Langmuir-Blodgett film [96]. Nevertheless, one key point in the development of biotechnological devices is the long term stability of enzymes. If these biological catalysts are very efficient *in vivo*, they often suffer from weak stability when extracted from their physiological environment. Enzyme encapsulation in silica-derived sol-gel materials has been demonstrated to stabilize many enzymes. This procedure was applied to hydrogenase [97]. The majority of hydrogenase was shown to be entrapped in the gel and protected against proteolysis. Hydrogenase/sol-gel pellets retained 60% of the specific mediated activity for H_2 production displayed by hydrogenase in solution. The gel-encapsulated enzyme retained its activity for long periods, *i.e.* 80% of the activity after four weeks at room temperature. Notably, by doping the hydrogenase-containing sol-gel materials with MWCNTs Zadvorny *et al.* demonstrated a 50% increase in dihydrogen production [98]. Furthermore stabilization of hydrogenase was proved through encapsulation process.

One alternative for green hydrogen production is to synthesize metal complexes that mimic the active site of enzymes. Huge work has been done in that field in order to obtain bioinspired models that could produce H_2 as efficiently as hydrogenase, while being much more stable [99]. The most performing complex involves mononuclear nickel diphosphine complex. This complex is inspired from the active sites of both [NiFe] and [FeFe] hydrogenases and displays remarkable catalytic proton reduction in organic solvent [100]. Le Goff *et al.* took benefice from this complex and from the results obtained by immobilization of hydrogenase on CNT networks [44]. The authors successfully immobilized the nickel complex onto carbon nanotube networks by covalent coupling [101]. Such construction was demonstrated to be very efficient for dihydrogen production in aqueous solution, evolving dihydrogen with overvoltage less than 20 mV and exceptional stability.

7. Carbon nanotubes for biofuel cells: an attractive green alternative

Beside researches towards decrease in chemical catalyst amount and discovery of less expensive catalysts (as alloys for example), a new concept emerged early in 1964 by Yahiro *et al.* [102]. A fuel cell was constructed using usual O_2 reduction at platinum modified electrode in the cell cathodic compartment, but using glucose as a fuel in the anodic compartment. The innovative idea was the use of an enzyme specific for fuel oxidation instead of platinum. For glucose oxidation, glucose oxidase was tested as the anodic catalyst. The fuel cell delivered 30 nA cm^{-2} at 330 mV…a very low power density indeed but the proof of concept of biofuel cell was born. Generally speaking these biofuel cells function as fuel cells but used enzymes instead of noble metals as catalysts (Figure 5). They are referred as enzymatic biofuel cells. Microorganisms can also be used as catalysts, defining microbial fuel cells. Microbial biofuel cells use the metabolism of microorganisms under anaerobic conditions to oxidize fuel [103-104]. Although a promising concept, little is known yet about the mechanisms by which fuel is oxidized at the anode. The involvement of nanowires, electron transfer mediators, either membrane-bound or excreted, is supposed to be responsible for the cell current. Enzymatic biofuel cells are however more efficient because no mass transfer limitations across the cell membrane exist.

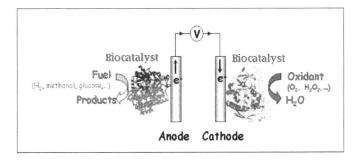

Figure 5. Schematic representation of an enzymatic biofuel cell.

The advantages of enzymatic biofuel cells over fuel cells are multiple. Biocatalysts are widespread, then *a priori* inexpensive, and biodegradable. Enzymes are highly efficient and specific to their substrates. The substrate specificity decreases reactant cross-over, and might theoretically allow to design fuel cells with no membrane between the anodic and cathodic compartments. Both costs are reduced and the design is simplified. A large variety of fuels and oxidants can be used to feed the biofuel cells, as opposed to the poor available fuels and oxidant in classical fuel cells (dihydrogen, methanol, oxygen). Indeed, many enzymes are nowadays characterized which differ by their natural abundant substrates. Dihydrogen, but also various inexpensive sugars can thus be used as efficient fuels at the anode. Furthermore, the involvement of cascades of enzymes can enhance the cell performance because of the summation of the electrons from each enzymatic reaction [105]. Finally, biofuel cells can deliver power under soft working conditions, as enzymes usually perform their enzymatic reactions at mild pH and temperature. Nevertheless, some extremophilic enzymes operate in extreme acidic or basic pH, as well as at high temperatures (around 90°C) or high pressure, offering the possibility to develop biofuel cell devices for special applications requiring extreme working conditions [106]. The applications of biofuel cells are still in their infancy. They are mainly thought to power small portable devices. Remarkable progress has been reported for implantable biofuel cells during the last year to power drug pumps, glucose sensors, vision devices [107-109].

The most common redox couple that has been used in biofuel cells is sugar/O_2, essentially because of sugar and O_2 abundance in nature and their essential role in living metabolism. In particular, glucose is an important metabolite and a source of energy for many living organisms. In that field, CNTs have been widely used, both at the anode and cathode. Glucose/O_2 biofuel cell is thus a very pertinent investigation field to investigate the role of CNTs. A view of some typical results is presented in Table 2.

Enzymes	Mediators	Power density	Ref
Anode / Cathode	Anode / Cathode	µW cm^{-2}	
Gox / Laccase	Ferrocene / -	15	[111]
GDH / BOD	PQQ / -	23	[82]
Gox / Pt	Ferrocenecarboxaldehyde / -	51	[112]
GDH/ BOD	Poly(brilliant cresyl blue) / -	54	[113]
GDH / laccase	Azine dies / -	58	[114]
Gox / Pt	Benzoquinone / -	77	[52]
Gox / Laccase	Ferrocene / ABTS	100	[115]
Gox / BOD	Ferrocene methanol / ABTS	120	[116]
GDH / Laccase	- / -	131	[117]
CDH / Pt	Os complex / -	157	[118]
Gox / Laccase	- / -	1300	[119]

Gox: Glucose oxidase; GDH: Glucose dehydrogenase; BOD: Bilirubin oxidase; ABTS: 2, 2'-azino-bis(3-ethylbenzothia-zoline-6-sulfonate) diammonium; CDH: cellobiose dehydrogenase.

Table 2. Performances of glucose/O_2 fuel cells.

Data highlight that kinetics of bioelectrochemical reactions, thus power density, largely depends on the experimental conditions, *i.e.* enzyme and mediators, T°, pH, concentration of substrate, electrolyte and type of electrode construction. Highest values are obtained with mediatorless fuel cells, reaching power densities upper than 1 mW cm^{-2} which is sufficient to power small electrical devices. It appears that direct connection of copper enzymes, namely laccase or BOD, for oxygen reduction at the cathode can be quite easily obtained with the help of CNT network. Direct connection of enzymes for glucose oxidation is conversely hardly observed, even on CNT coatings. From literature examination direct connection of

Gox at electrode interfaces is still controversial. Due to the peculiar structure of Gox, a dimer with flavin adenine dinucleotide active site buried within a thick and isolated protein shell, it is understandable that electrical connection of Gox could be unexpected. A recent work concluded that CNTs were capable to electrically connect Gox, but this connection was unfruitful for glucose catalytic oxidation [110].

8. Carbon nanotubes for bioelectrooxidation of H_2: towards H_2/O_2 biofuel cells

We already described above hydrogenases, the enzymes that convert with high specificity and efficiency protons into dihydrogen. Most of these biocatalysts are also efficient in the oxidation of dihydrogen into protons. Consequently this allows to imagine biofuel cells in which the fuel would be dihydrogen, exactly as in PEM fuel cells. As hydrogenases are able to oxidize dihydrogen with very low overvoltage, the open circuit voltage for the biofuel cell using oxygen at the cathode, is expected to be not far from the thermodynamic one, *i.e.* 1.23 V. Hence, high power densities are expected, provided that a strong and efficient electrical connection between hydrogenase and electrode can be achieved. Simple adsorption of hydrogenase was performed in a first step, because it allowed a direct oxidation of dihydrogen without any redox mediators [120]. Catalytic mechanisms associated with dihydrogen oxidation at the active site were largely studied. The effect of strong hydrogenase inhibitors such as oxygen and CO were explored by this mean, leading to nice developments in engineering of more tolerant hydrogenases [121] or use of naturally resistant hydrogenases [122, 123]. However, this immobilization procedure relies on a monolayer of enzyme, which furthermore suffers from quick desorption. Otherwise, multilayer enzymatic films require a redox mediator so that even the last layer far from the electrode could be connected. Other immobilization processes are thus needed, that can favor an enhancement in both the amount of connected hydrogenases as well as their stability, while preserving their functionality.

Carbon nanotube networks constituted technological breakthroughs in that way. All the recent developments using immobilization of hydrogenases onto carbon nanotubes point out improved catalytic currents essentially related to an increase in the active area of the electrode. The respective role of metallic-SWCNTs against semiconducting one was explored for dihydrogen oxidation by immobilized hydrogenase [124]. A higher oxidation process was revealed when the nanotube mixture was enriched in metallic SWCNT. The study furthermore suggested no need of oxygenated SWCNTs for efficient anchoring of hydrogenases. The catalytic current enhancement was claimed to be due to an increase in active electrode surface area and an improved electronic coupling between hydrogenase redox active sites and the electrode surface. In most cases, however, CNTs are used as a mixture of metallic and semi-conducting tubes. Oxidation of the mixture yields the defects and functionalities described above in this review. Advantage is gained due to these chemical functions quite easily generated on the surface of the carbon nanotubes. Electrodes modified by carbon nanotubes are thus expected to offer numerous anchoring sites for stable hydrogenase immobilization. The literature provides a few examples of efficient immobilization of hydrogenase on carbon nanotubes coatings bearing various functionalities. Both SWCNTs and MWCNTs are used. Notably, more and more articles are devoted nowadays to this domain in hydrogenase research. A bionanocomposite made of the hydrogenase, MWCNTs and a thiopyridine derivative was proved to form stable monolayers when transferred by Langmuir-Blodgett method on indium tin oxide electrode surfaces [125]. A greater amount of electroactive hydrogenase towards dihydrogen oxidation was demonstrated to be adsorbed on the Langmuir-Blodgett films. De Lacey and co-workers grew MWCNTs on electrode by chemical vapor deposition of acetylene [65]. A high density of vertically aligned carbon nanotubes was obtained, which were functionalized by electroreduction of a diazonium salt for covalent binding of hydrogenase. High coverage of electroactive enzyme was measured, suggesting that almost all the functionalized CNT surface was accessible to hydrogenase. Great stabilization of the catalytic current for H_2 oxidation was obtained, with no decrease in current density after one month. Another work by Heering and coworkers studied a gold electrode pre-treated by polymyxin then a multilayer of carbon nanotubes [126]. Polymyxin was shown to help in the stable attachment of hydrogenase on the gold electrode. Using adsorption of hydrogenase on a nanotube layer pretreated with polymyxin the current density for H_2 oxidation was an order of magnitude higher than at the gold electrode only modified by polymixin. This result was supposed to origin from greater surface area even though only the top of the nanotube layer was supposed to be accessible to the enzyme. The catalytic current was stable with time, at least for two hours under continuous cycling, and several days upon storage under ambient conditions. AFM visualization of hydrogenase immobilized onto polymyxin-treated SWCNT layer on SiO_2 revealed that hydrogenase was structurally intact and preferentially adsorbed on the sidewalls of the CNTs rather than on SiO_2 [126].

In our laboratory, we immobilized the [NiFe] hydrogenase from a mesophilic anaerobic bacterium (the sulfate reducing bacterium *Desulfovibrio fructosovorans* Df) by adsorption onto SWCNT films [44]. The current for direct H_2 oxidation was shown to increase with the amount of SWCNTs in the coating (Figure 6).

Because non-turnover signals were not detected for hydrogenase in these conditions, the increase in surface area was evaluated using a redox protein as a probe. It was shown that SWCNTs induced one order larger surface area. The same hydrogenase was entrapped in methylviologen functionalized polypyrrole films coated onto SWCNTs and MWCNTs [127]. Although no direct electrical hydrogenase connection was observed, an efficient dihydrogen oxidation through a mediated process occurred. It was concluded that the entrapment of hydrogenase into the redox polymer coated onto CNTs combined the electron carrier properties of redox probes, the flexibility of polypyrroles, and the high electroactive area developed by CNTs. The reason why no direct connection could be observed is however not clearly understood yet. In our group we handled immobilization of hydrogenase on a film obtained by electropolymerization of a phenothiazine dye on a SWCNT coating [81]. The phenothiazine dye was shown to be able to mediate dihydrogen oxidation but also to serve as an anchor for the enzyme when adsorbed or when electropolymerized. Higher current density than in the absence of SWCNT was observed. In addition, a wider potential window for dihydrogen oxidation was reached as well as very stable electrochemical signals with time. We postulated that the conductive polymer which was electropolymerized onto CNTs could play a multiple role: enhancement of the electroactive surface area, enhancement of redox mediator units due to phenothiazine monomers entrapped in the polymer matrix, enhancement of hydrogenase anchorage sites. We have already mentioned in this review the advantages of a direct electron transfer over a mediated one for H_2 oxidation, including gain in over-potential values, less interferences due to enzyme specificity, absence of redox mediators that could be difficult to co-immobilize with the enzyme... Functionalized carbon nanotube films were evaluated in our group as platforms for various hydrogenases, that present a very different environment of FeS cluster electron relay. Dihydrogen oxidation was studied at gold electrodes modified with functionalized self-assembled-monolayers [128]. As expected, dihydrogen oxidation process was demonstrated to be driven by electrostatic or hydrophobic interactions according to the specific environment of the surface electron relay. Interestingly, at CNT coatings, although CNTs were negatively charged, direct electrical connection of hydrogenases that present a negatively charged patch around the FeS surface electron relay was observed [44, 123]. In other words, despite unfavourable electrostatic interactions, direct electron transfer process for dihydrogen oxidation was achieved. One important conclusion was that on such CNT films, the nanometric size of the CNTs allows a population of hydrogenases to be directly connected to a neighbouring nanotube, hence allowing direct electron transfer for H_2 oxidation, whatever the orientation of the enzyme.

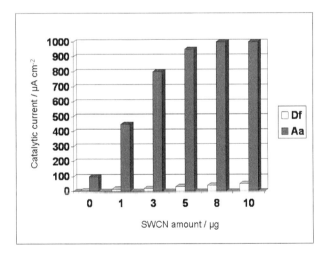

Figure 6. Comparative evolution of the catalytic current for dihydrogen oxidation with the amount of SWCNTs deposited at a graphite electrode in the case of hydrogenases from *Aquifex aeolicus* (Aa) or *Desulfovibrio fructosovorans* (Df). Catalytic currents are measured using voltammetry under H_2 at 60 and 25°C for Aa and Df respectively.

However, the extreme oxygen sensitivity of hydrogenases used in the former studies yielded an intensive research towards more resistant enzymes. During the last years, four [NiFe] membrane-bound hydrogenases have been discovered from aerobic or extremophilic organisms [128, 129-132]. They have been demonstrated to oxidize H_2 in the presence of oxygen and CO. The crystallographic structure of three of them has been resolved, showing that an uncommon [4Fe-3S] cluster proximal to the active site prevents deleterious oxygen attack. Of course, the sensitivity to oxygen, and also to CO, of most hydrogenases known before was a strong limitation for their potential use in biotechnological devices. Therefore these resistant biocatalysts open new avenues towards a biohydrogen economy. No doubt that these researches will increase in the next future. To date, two main studies report the immobilization of resistant hydrogenase on CNT-modified electrodes. Krishnan *et al.* very recently modified MWCNTs by pyrenebutyric acid, and demonstrated it was an efficient platform for stable O_2-resistant hydrogenase linkage [133]. In our group, original use of a hyperthermophilic O_2- and CO-resistant hydrogenase allowed the increase in the catalytic current for direct H_2 oxidation on a large range of temperature up to 70°C. Attempts to enhance the number of electrically connected hydrogenase succeeded by use of coatings of chemically oxidized SWCNTs [123]. Values as high as 1 mA cm^{-2} were reached depending on the amount of SWCNTs used in the coating (Figure 6). For the lowest amounts of SWCNTs, the increase in the catalytic current was demonstrated to be essentially due to the increase in surface area. However the catalytic current rapidly reached a plateau, although the peak current for the redox probe still increased, suggesting rapid saturation of the surface.

9. Design of a H_2/O_2 biofuel cell based on carbon nanotubes-modified electrodes

H_2/O_2 biofuel cells did not get much attention before O_2 and CO resistant hydrogenases were proved to be efficient for H_2 oxidation when immobilized onto electrode surfaces. Even though more and more efficient hydrogenase immobilization procedures are nowadays reported, few H_2/O_2 biofuel cells are described. An early study by Armstrong's group in 2006 [134] demonstrated that simple adsorption on graphite electrode of hydrogenase at the anode and laccase (a copper protein for O_2 reduction) at the cathode, allowed a wristwatch to run for 24h. Power density of around 5 μW cm^{-2} at 500 mV was delivered with no membrane between the two compartments providing hydrogenase was extracted from *Ralstonia metallireducens*. As this is an aerobic bacterium, the result underlined that the H_2/O_2 biofuel cell could operate only with O_2 resistant hydrogenase. In 2010, the same group improved the device by using another O_2 resistant hydrogenase from *Escherichia coli* and bilirubin oxidase (BOD), another copper protein more efficient than laccase towards oxygen reduction because being able to function at neutral pH [135]. The oxygen reductase was covalently linked to the graphite electrode which had been modified by diazonium salt reduction. The power density was enhanced compared to the former study reaching 63 μW cm^{-2}. But most of all, this work provided a nice understanding of the operating conditions of such H_2/O_2 fuel cells involving hydrogenase as anode catalyst.

Due to the understanding of how hydrogenases could be efficiently connected at CNT-coated electrodes, a huge step jumped over very recently. First, using covalent attachment of both O_2 resistant hydrogenase and BOD on pyrene derivative functionalized MWCNTs, a membrane-less biofuel cell was designed fed with a non-explosive 80/20 dihydrogen/air mixture [133]. This biofuel cell displayed quite a good stability with time and a much higher power density than reported before. Indeed, an average power density of 119 μW cm^{-2} was measured. Low solubility of oxygen and weak affinity of BOD for oxygen was shown to limit the cathodic current. Secondly in our group, a more performant H_2/O_2 mediatorless biofuel cell was constructed based on one step covalent attachment directly on SWCNTs of an hyperthermophilic O_2 resistant hydrogenase at the anode and BOD at the cathode [136] (Figure 7).

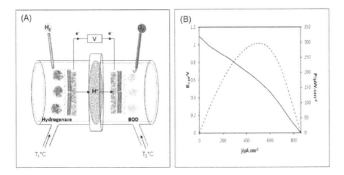

Figure 7. (A) Schematic representation of H_2/O_2 biofuel cell with O_2 resistant hydrogenase at the anode and bilirubin oxidase (BOD) at the cathode. Each half cell, separated by a Nafion® membrane, is independently thermoregulated with waterbaths. (B) Performance of the H_2/O_2 biofuel cell.

Taking advantage of temperature, the biofuel cell delivered power densities up to 300 μW cm^{-2} at 0.6V with an OCV of 1.1V, which is the highest performance ever reported. Furthermore, promising stability of the biofuel cell during 24h of continuous use lets us consider this device as an alternative power supply for small portable applications. The analysis of the fuel cell parameters during polarization, allows us to define the potential window in which the fuel cell fully operates. Interestingly, in Armstrong's group [135] and in our group, different approaches on the settings of biofuel cell working conditions, led to similar observations of an unexpected increasing anodic potential. This high oxidizing potential generates an inactive state of hydrogenase active site. It is worth noticing that this hydrogenase inactivation occurred under anaerobic conditions in our group while it was under aerobic conditions in Armstrong's group. Consequently, dramatic loss in power densities was observed. By applying negative potential to the anode, and thus providing electrons to the active site, we were unable to reactivate hydrogenase. Another protocol used by Armstrong, consisted to add a second hydrogenase coated anode, unconnected to the system but present in the anodic half-cell which was consequently unaffected by the oxidizing potential but still in presence of O_2. This second anode, under H_2 oxidation was used as an electron supplier and connected to the first electrode. This procedure reactivated hydrogenase and allowed full recovery of OCV. It is of relevant interest to overcome hydrogenase inactivation in H_2/O_2 biofuel cell.

10. Future directions

As reviewed in this chapter, many of the CNTs based technologies are promising for the development of a green hydrogen economy. Not only abiotic dihydrogen storage, but also microbial dihydrogen production and use of this green dihydrogen in biofuel cells can take advantages of the outstanding properties of CNTs. In all these applications, CNTs appear to play multiple roles including increase in surface area, increase in electron transfer rate, increase in directly connected enzymes. Possible protection against oxygen damage of enzymes has even been strongly suggested. Use of CNTs thus allows to architecture three-dimensional nanostructured interfaces which can be an alternative to strictly orientated proteins or enzymes for high direct electron transfer interfacial processes. The ease in obtaining tuned surface functionalizations is one of the very attracting points in view of the development of efficient bioelectrodes.

This is in particular the case for biofuel cells using dihydrogen as a fuel. During the last years, tremendous research on hydrogenase, the key enzyme for dihydrogen conversion, has led to the discovery, then control of some hydrogenases presenting properties that allow their use in biotechnological devices. During this year, based on these new resistant enzymes and on improved knowledge of how CNTs can enhance direct current densities, two H_2/O_2 biofuel cells have been reported. Although these biofuel cells constitute the first device using hydrogenases, they already deliver sufficient power density for small portable applications. No doubt that this research field will gain more and more interest in a next future.

However, various directions might be followed to further improve the biological system in such a way it could be commercially available. One is the enhancement of long-term stability of the device, which is obviously the critical point shared by (bio)fuel cells, yet. Search for more stable enzymes in the biodiversity or enzyme engineering has to be explored. Protection of enzymes by various encapsulation procedures could be another solution given efficient interfacial electron transfer can be reached. The use of whole microorganisms with controlled and driven metabolism, or at least immobilization of naturally encapsulated enzymes will be a next step. As an example, reconstitution of proteoliposomes with a membrane-bound hydrogenase was proved to enhance the stability of the enzyme [137]. This could be a novel route for preserving enzymes in their physiological environment, hence enhancing their stability. New enzymes, with outstanding properties (T°, pH, inhibitors, substrate affinity…) have to be discovered and studied. Notably, two very recent publications report on a new thermostable bilirubin oxidase and a tyrosinase which present outstanding resistances to serum constituents [138, 139]. These two new enzymes appear to be able to efficiently replace the currently used BOD for implantable applications of biofuel cells.

More sophisticated materials interfaces, constituted of mixtures of CNTs with other conducting materials could bring a hierarchical porosity necessary for both enzyme immobilization and substrate diffusion. Carbon fibers, mesoporous carbon templates could be used to build very interesting new electrochemical interfaces. This diversity in potential carbon materials for efficient enzyme immobilization would be a key step to go through the difficulties linked to CNTs, *i.e.* effective cost for separation and purification as well as possible toxicity. Finally, to avoid the membrane between the cathodic and anodic compartments, and build a miniaturized biofuel cell, unusual cell designs, such as microfluidic or flow-through systems, are likely to open new avenues. All these future developments will certainly require a multidisciplinary approach, coupling electrochemists with biochemists and physicists, and coupling methods such as electrochemistry and spectrometry, electrochemistry and molecular genetics or electrochemistry and materials chemistry. This multidisciplinary willingness will help in the elucidation of the interactions between enzymes and nanostructured materials at the nanoscale and yield innovative nanobiotechnological approaches and applications.

Acknowledgements

We gratefully acknowledge the contribution of Marielle Bauzan (Fermentation Plant Unit, IMM, CNRS, Marseille, France) for growing the bacteria, Dr Marianne Guiral, Dr Marianne Ilbert and Pascale Infossi for fruitful discussions. This work was supported by research grants from CNRS, Région PACA and ANR.

Author details

Anne De Poulpiquet, Alexandre Ciaccafava, Saïda Benomar,
Marie-Thérèse Giudici-Orticoni and Elisabeth Lojou*

*Address all correspondence to: lojou@imm.cnrs.fr

Bioénergétique et Ingénierie des Protéines, CNRS - AMU - Institut de Microbiologie de la Méditerranée, France

References

[1] Verne, J. (1874). L'île mystérieuse.

[2] Grove, W. (1838). On a new voltaic combination. *Philosophical Magazine and Journal of Science*, 13, 430.

[3] Charlou, J. L., Donval, J. P., Konn, C., Ondréas, H., Fouquet, Y., Jean-Baptiste, P., & Fourré, E. (2010). High production and fluxes of H$_2$ and CH$_4$ and evidence of abiotic hydrocarbon synthesis by serpentinization on ultramafic-hosted hydrothermal systems on Mid-Atlantic Ridge. Rona P., Devey C., Dyment J. Murton B. Editors,. *"Di-*

versity of hydrothermal systems on slow spreading ocean ridges" Edited by AGU Geophysical *monograph series*, 188, 265-296.

[4] Oberlin, A., Endo, M., & Koyama, T. (1976). Filamentous growth of carbon through benzene decomposition. *Journal of Crystal Growth*, 32(3), 335-349.

[5] Iijima, S. (1991). Helical microtubules of graphitic carbon. *Nature*, 354-56.

[6] Saito, R., Dresselhaus, G., & Dresselhaus, M. (1998). Physical Properties of Carbon Nanotubes. *Imperial College Press, London.*

[7] Kirk, Othmer. (1996). Encyclopedia of Chemical Technology, (4th ed.). *Wiley and Sons, New York.*

[8] Heitner-Wirguin, C. (1996). Recent advances in perfluorinated ionomer membranes: structure, properties and applications. *Journal of Membrane Science*, 120(1), 1-33.

[9] Cracknell, J., Vincent, K., & Armstrong, F. (2008). Enzymes as working or inspirational electrocatalysts for fuel cells and electrolysis. *Chemical Review*, 108(7), 2439-2461.

[10] Guiral, M., Prunetti, L., Aussignargues, C., Ciaccafava, A., Infossi, P., Ilbert, M., Lojou, E., & Giudici-Orticoni, M. T. (2012). The hyperthermophilic bacterium *Aquifex aeolicus*: from respiratory pathways to extremely resistant enzymes and biotechnological applications. *Advances in Microbiological Physiology; to be edited.*

[11] Ivanov, I., Vidakovic-Koch, T., & Sundmacher, K. (2010). Recent Advances in Enzymatic Fuel Cells:. *Experiments and Modeling Energies*, 3(4), 803-846.

[12] Tran, P., Artero, V., & Fontecave, M. (2010). Water electrolysis and photoelectrolysis on electrodes engineered using biological and bio-inspired molecular systems. *Energy & Environmental Science*, 3(6), 727-747.

[13] Lojou, E. (2011). Hydrogenases as catalysts for fuel cells: Strategies for efficient immobilization at electrode interfaces. *Electrochimica Acta*, 56(28), 10385-10397.

[14] Schnorr, J. M., & Swager, T. M. (2011). Emerging Applications of Carbon Nanotubes. *Chemistry of Matererials*, 23(3), 646-657.

[15] Iijima, S., & Ichihashi, T. (1993). Single-shell carbon nanotubes of 1-nm diameter. *Nature*, 363, 603-605.

[16] Dresselhaus, M., Dresselhaus, G., & Jorio, A. (2004). Unusual properties and structure of carbon nanotubes. *Annual Review of Material Research*, 34-247.

[17] Botzung, M., Chaudourne, S., Gillia, O., Perret, C., Latroche, M., Percheron-Guegan, A., & Marty, P. (2008). Marty Simulation and experimental validation of a hydrogen storage tank with metal hydrides. *International Journal of Hydrogen Energy*, 33(1), 98-104.

[18] Vajo, J., Li, W., & Liu, P. (2010). Thermodynamic and kinetic destabilization in $LiBH_4/Mg_2NiH_4$: promise for borohydride-based hydrogen storage. *Chemical Communications*, 46(36), 6687-6689.

[19] Li, C., Peng, P., Zhou, D., & Wan, L. (2011). Research progress in LiBH$_4$ for hydrogen storage: A review. *International Journal of Hydrogen Energy*, 36(22), 14512-14526.

[20] Yao, X., Wu, C., Du, A., Lu, G., Cheng, H., Smith, S., Zou, J., & He, Y. (2006). Mg-based nanocomposites with high capacity and fast kinetics for hydrogen storage. *Journal of Physical Chemistry B*, 110(24), 11697-11703.

[21] Luo, Y., Wang, P., Ma , L. P, & Cheng, M. (2007). Enhanced hydrogen storage properties of MgH$_2$ co-catalyzed with NbF$_5$ and single-walled carbon nanotubes. *Scripta Materialia*, 56(9), 765-768.

[22] Wu, C., Wang, P., Yao, X., Liu, C., Chen, D., Lu, G., & Cheng, H. (2006). Effect of carbon/noncarbon addition on hydrogen storage behaviour of magnesium hydride. *Journal of Alloys Compounds*, 414-259.

[23] Dillon, A., Jones, K., Bekkedahl, T., Kiang, C., Bethune, D., & Heben, M. (1997). Storage of hydrogen in single-walled carbon nanotubes. *Nature*, 386(6623), 377-379.

[24] Ding, F., & Yakobson, B. (2011). Challenges in hydrogen adsorptions: from physisorption to chemisorption. *Frontiers of Physics*, 6(2), 142-150.

[25] Nikitin, A., Li, X., Zhang, Z., Ogasawara, H., Dai, H., & Nilsson, A. (2008). Hydrogen storage in carbon nanotubes through the formation of stable C-H bonds. *Nano Letters*, 8(1), 162-167.

[26] Sahaym, U., & Norton, M. (2008). Advances in the application of nanotechnology in enabling a hydrogen economy. *Journal of Material Sciences*, 43(16), 5395-549.

[27] Tang, Z., Poh, C., Lee, K. K., Tian, Z., Chua, D., & Lin, J. (2010). Enhanced catalytic properties from platinum nanodots covered carbon nanotubes for proton-exchange membrane fuel cells. *Journal of Power Sources*, 195(1), 155-159.

[28] Lin, J., Kamavaram, V., & Kannan, A. (2010). Synthesis and characterization of carbon nanotubes supported platinum nanocatalyst for proton exchange membrane fuel cells. *Journal of Power Sources*, 195(2), 466-470.

[29] Kongkanand, A, Kuwabata, S, Girishkumar, G, & Kamat, P. (2006). Single-wall carbon nanotubes supported platinum nanoparticles with improved electrocatalytic activity for oxygen reduction reaction. *Langmuir*, 22(5), 2392-2396.

[30] Lin, J., Mason, C., Adame, A., Liu, X., Peng, X., & Kannan, A. (2010). Synthesis of Pt nanocatalyst with micelle-encapsulated multi-walled carbon nanotubes as support for proton exchange membrane fuel cells. *Electrochimica Acta*, 55(22), 6496-6500.

[31] Lee, C., Ju, Y. C., Chou, P. T., Huang, Y. C., Kuo, L. C., & Oung, J. C. (2005). Preparation of Pt nanoparticles on carbon nanotubes and graphite nanofibers via self-regulated reduction of surfactants and their application as electrochemical catalyst. *Electrochemistry Communications*, 7(4), 453-458.

[32] He, D., Zeng, C., Xu, C., Cheng, N., Li, H., & Mu, S. (2011). Polyaniline-Functionalized Carbon Nanotube Supported Platinum Catalysts. *Langmuir*, 27(9), 5582-5588.

[33] Ludwig, R., Harreither, W., Tasca, F., & Gorton, L. (2011). Cellobiose Dehydrogenase: A Versatile Catalyst for Electrochemical Applications. *ChemPhysChem.*, 11(13), 2674-2697.

[34] Cang-Rong, J., & Pastorin, G. (2009). The influence of carbon nanotubes on enzyme activity and structure: investigation of different immobilization procedures through enzyme kinetics and circular dichroism studies. *Nanotechnology*, 20(25), 255102.

[35] Matsuura, K., Saito, T., Okasaki, T., Oshima, S., Yumura, M., & Iijima, S. (2006). Selectivity of water-soluble proteins in single-walled carbon nanotube dispersions. *Chemical Physics Letters*, 429(4-6), 497-502.

[36] Patolsky, F., Weizmann, Y., & Willner, I. (2004). Long-range electrical contacting of redox enzymes by SWCNT connectors. *Angewandte Chemistry International Edition*, 43(16), 2113-2117.

[37] Liu, J., Chou, A., Rahmat, W., Paddon-Row, M., & Gooding, J. (2005). Achieving direct electrical connection to glucose oxidase using aligned single walled carbon nanotube arrays. *Electroanalysis*, 17(1), 38-46.

[38] Wang, J. (2005). Carbon-nanotube based electrochemical biosensors: A review. *Electroanalysis*, 17(1), 7-14.

[39] Wildgoose, G., Banks, C., Leventis, H., & Compton, R. (2006). Chemically modified carbon nanotubes for use in electroanalysis. *Microchimica Acta*, 152(3-4), 187-214.

[40] Dumitrescu, I., Unwin, P., & Macpherson, J. (2009). Electrochemistry at carbon nanotubes: perspective and issues. *Chemical Communications*, 45, 6886-4901.

[41] Ji, P., Tan, H., Xu, X., & Feng, W. (2010). Lipase Covalently Attached to Multiwalled Carbon Nanotubes as an Efficient Catalyst in Organic Solvent. *AIChE Journal*, 56(11), 3005-3011.

[42] Upadhyayula, V., & Gadhamshetty, V. (2010). Appreciating the role of carbon nanotube composites in preventing biofouling and promoting biofilms on material surfaces in environmental engineering: A review. *Biotechnological Advances*, 28(6), 802-816.

[43] Lojou, E., Luciano, P., Nitsche, S., & Bianco, P. (1999). Poly(ester-sulfonic acid):modified carbon electrodes for the electrochemical study of c-type cytochromes. *Electrochimica Acta*, 44(19), 3341-3352.

[44] Lojou, E., Luo, X., Brugna, M., Candoni, N., Dementin, S., & Giudici-Orticoni, M. T. (2008). Biocatalysts for fuel cells: efficient hydrogenase orientation for H_2 oxidation at electrodes modified with carbon nanotubes. *Journal of Biological Inorganic Chemistry*, 13(7), 1157-1167.

[45] Minteer, S., Atanassov, P., Luckarift, H., & Johnson, G. (2012). New materials for biological fuel cells. *Material Today*, 15(4), 166-173.

[46] Weigel, M., Tritscher, E., & Lisdat, F. (2007). Direct electrochemical conversion of bi-lirubin oxidase at carbon nanotube-modified glassy carbon electrodes. *Electrochemistry Communications*, 9(4), 689-693.

[47] Pumera, M., & Smid, B. (2007). Redox protein noncovalent functionalization of dou-ble-wall carbon nanotubes: Electrochemical binder-less glucose biosensor. *Journal of Nanosciences and Nanotechnology*, 7(10), 3590-3595.

[48] Willner, I., Yan, Y. M., Willner, B., & Tel-Vered, R. (2009). Integrated Enzyme-Based Biofuel Cells-A Review. *Fuel Cells*, 09(1), 7-24.

[49] Ueda, A., Kato, D., Kurita, R., Kamata, T., Inokuchi, H., Umemura, S., Hirono, S., & Niwa, O. (2011). Efficient Direct Electron Transfer with Enzyme on a Nanostructured Carbon Film Fabricated with a Maskless Top-Down UV/Ozone Process. *Journal of the American Chemical Society*, 133(13), 4840-4846.

[50] Tasca, F., Gorton, L., Harreither, W., Haltrich, D., Ludwig, R., & Nöll, G. (2088). Di-rect electron transfer at cellobiose dehydrogenase modified anodes for biofuel cells. *Journal of Physical Chemistry C*, 112(26), 9956-9961.

[51] Zheng, W., Zhao, H., Zhou, H., Xu, X., Ding, M., & Zheng, Y. (2010). Electrochemis-try of bilirubin oxidase at carbon nanotubes. *Journal of Solid State Electrochemistry*, 14(2), 249-254.

[52] Zheng, W., Zhou, H., Zheng, Y., & Wang, N. (2008). A comparative study on electro-chemistry of laccase at two kinds of carbon nanotubes and its application for biofuel cell. *Chemical Physics Letters*, 381-385.

[53] Zhao, L., Liu, H., & Hu, N. (2006). Assembly of layer-by-layer films of heme proteins and single-walled carbon nanotubes: electrochemistry and electrocatalysis. *Analytical Bioanalytical Chemistry*, 384(2), 414-422.

[54] Liu, G., & Lin, Y. (2006). Amperometric glucose biosensor based on self-assembling glucose oxidase on carbon nanotubes. *Electrochemistry Communications*, 8(2), 251-256.

[55] Iost, R., & Crespilho, F. (2012). Layer-by-layer self-assembly and electrochemistry: Applications in biosensing and bioelectronics. *Biosensors Bioelectronics*, 31(1), 1-10.

[56] Gooding, J., Wibowo, R., Liu, J., Yang, W., Losic, D., Orbons, S., Mearns, F., Shapter, J., & Hibbert, D. (2003). Protein electrochemistry using aligned carbon nanotube ar-rays. *Journal of the American Chemical Society*, 125(30), 9006-9007.

[57] Yu, X., Chattopadhyay, D., Galeska, I., Papadimitrakopoulos, F., & Rusling, J. (2003). Peroxidase activity of enzymes bound to the ends of single-wall carbon nanotube for-est electrodes. *Electrochemistry Communications*, 5(5), 408-411.

[58] Esplandiu, M., Pacios, M., Cyganek, L., Bartroli, J., & Del Valle, M. (2009). Enhancing the electrochemical response of myoglobin with carbon nanotube electrodes. *Nano-technology*.

[59] Santhosh, P., Gopalan, A., & Lee, K. (2006). Gold nanoparticles dispersed polyaniline grafted multiwall carbon nanotubes as newer electrocatalysts: Preparation and performances for methanol oxidation. *Journal of Catalysis*, 238(1), 177-185.

[60] Nazaruk, E., Karaskiewicz, M., Zelechowska, K., Biernat, J., Rogalski, J., & Bilewicz, R. (2012). Powerful connection of laccase and carbon nanotubes Material for mediator-free electron transport on the enzymatic cathode of the biobattery. *Electrochemistry Communications*, 14(1), 67-70.

[61] Sadowska, K., Stolarczyk, K., Biernat, J., Roberts, K., Rogalski, J., & Bilewicz, R. (2010). Derivatization of single-walled carbon nanotubes with redox mediator for biocatalytic oxygen electrodes. *Bioelectrochemistry*, 80(1), 73-80.

[62] Jeykumari, D., & Narayanan, S. (2008). Fabrication of bienzyme nanobiocomposite electrode using functionalized carbon nanotubes for biosensing applications. *Biosensors Bioelectronics*, 23(11), 1686-1693.

[63] Wang, Z., Li, M., Su, P., Zhang, Y., Shen, Y., Han, D., Ivaska, A., & Niu, L. (2008). Direct electron transfer of horseradish peroxidase and its electrocatalysis based on carbon nanotube/thionine/gold composites. *Electrochemistry Communications*, 10(2), 306-310.

[64] Le Floch, F., Thuaire, A., Bidan, G., & Simonato, J. P. (2009). The electrochemical signature of functionalized single-walled carbon nanotubes bearing electroactive groups. *Nanotechnology*, 20(14), 45705.

[65] Alonso-Lomillo, M., Rüdiger, O., Maroto-Valiente, A., Velez, M., Rodriguez-Ramos, I., Munoz, F., Fernandez, V., & De Lacey, A. (2007). Hydrogenase-coated carbon nanotubes for efficient H_2 oxidation. *Nano Letters*, 7(6), 1603-1608.

[66] Jönsson-Niedziolka, M., Kaminska, A., & Opallo, M. (2010). Pyrene-functionalised single-walled carbon nanotubes for mediatorless dioxygen bioelectrocatalysis. *Electrochimica Acta*, 55(28), 8744-8750.

[67] Karachevtsev, V., Stepanian, S., Glamazda, A., Karachevtsev, M., Eremenko, V., Lytvyn, O., & Adamowicz, L. (2011). Noncovalent Interaction of Single-Walled Carbon Nanotubes with 1-Pyrenebutanoic Acid Succinimide Ester and Glucose oxidase. *Journal of Physical Chemistry C*, 115(43), 21072-21082.

[68] Lau, C, Adkins, E, Ramasamy, R, Lackarift, H, Johnson, G, & Atanassov, P. (2012). Design of Carbon Nanotube-Based Gas-Diffusion Cathode for O_2 Reduction by Multicopper Oxidases. *Advanced Energy Materials*, 2(1), 162-168.

[69] Xu, H., Xiong-Y, H., Zeng-X, Q., Jia, L., Wang, Y., & Wang-F, S. (2009). Direct electrochemistry and electrocatalysis of heme proteins immobilized in single-wall carbon nanotubes-surfactant films in room temperature ionic liquids. *Electrochemistry Communications*, 11(2), 286-289.

[70] Yan, Y., Zheng, W., Zhang, M., Wang, L., Su, L., & Mao, L. (2005). Bioelectrochemically functional nanohybrids through co-assembling of proteins and surfactants onto

carbon nanotubes: Facilitated electron transfer of assembled proteins with enhanced faradic response. *Langmuir*, 21(14), 65606566.

[71] Cosnier, S., Ionescu, R., & Holzinger, M. (2008). Aqueous dispersions of SWCNTs using pyrrolic surfactants for the electro-generation of homogeneous nanotube composites. Application to the design of an amperometric biosensor. *Journal of Materials Chemistry*, 18(42), 5129-5133.

[72] Gao, M., Dai, L., & Wallace, G. (2003). Biosensors based on aligned carbon nanotubes coated with inherently conducting polymers. *Electroanalysis*, 15(13), 1089-1094.

[73] Tsai, C. Y., Li, C. S., & Liao, W. S. (2006). Electrodeposition of polypyrrole-multiwalled carbon nanotube-glucose oxidase nanobiocomposite film for the detection of glucose. *Biosensors Bioelectronics*, 22(4), 495-500.

[74] Chen, H, & Dong, S. (2007). Direct electrochemistry and electrocatalysis of horseradish peroxidase immobilized in sol-gel-derived ceramic-carbon nanotube nanocomposite film. *Biosensors Bioelectronics*, 22(8), 1811-1815.

[75] Heller, A. (2006). Electron-conducting redox hydrogels: design, characteristics and synthesis. *Current Opinion in Chemical Biology*, 10(6), 664-672.

[76] Timur, S, Anik, U, Odaci, D, & Gorton, L. (2007). Development of a microbial biosensor based on carbon nanotube (CNT) modified electrodes. *Electrochemistry Communications*, 9(7), 1810-1815.

[77] Song, J., Shin, H., & Kang, C. (2011). A Carbon Nanotube Layered Electrode for the Construction of the Wired Bilirubin Oxidase Oxygen Cathode. *Electroanalysis*, 23(12), 2941-2948.

[78] Tiwari, I., & Singh, M. (2011). Preparation and characterization of methylene blue-SDS- multiwalled carbon nanotubes nanocomposite for the detection of hydrogen peroxide. *Microchimica Acta*, 174(3-4), 223-230.

[79] Pakapongpan, S., Palangsuntikul, R., & Surareungchai, W. (2011). Electrochemical sensors for hemoglobin and myoglobin detection based on methylene blue-multiwalled carbon nanotubes nanohybrid-modified glassy carbon electrode. *Electrochimica Acta*, 56(19), 6831-6836.

[80] Hoshino, T., Sekiguchi, S., & Muguruma, H. (2012). Amperometric biosensor based on multilayer containing carbon nanotube, plasma-polymerized film, electron transfer mediator phenothiazine, and glucose dehydrogenase. *Bioelectrochemistry*, 84-1.

[81] Ciaccafava, A., Infossi, P., Giudici-Orticoni, M. T., & Lojou, E. (2010). Stabilization role of a phenothiazine derivative on the electrocatalytic oxidation of hydrogen via *Aquifex aeolicus* hydrogenase at graphite membrane electrodes. *Langmuir*, 26(23), 18534-18541.

[82] Tanne, C., Göbel, G., & Lisdat, F. (2010). Development of a (PQQ)-GDH-anode based on MWCNT-modified gold and its application in a glucose/O_2-biofuel cell. *Biosensors Bioelectronics*, 26(2), 530-535.

[83] Reisner, E. (2011). Solar Hydrogen Evolution with Hydrogenases: From Natural to Hybrid Systems European. *Journal of Inorganic Chemistry* [7], 1005-1016.

[84] Vignais, P., & Billoud, B. (2007). Occurrence, classification, and biological function of hydrogenases: An overview. *Chemical Reviews*, 107(10), 4206-4272.

[85] Parkin, A., Cavazza, C., Fontecilla-Camp, J., & Armstrong, F. (2006). Electrochemical investigations of the interconversions between catalytic and inhibited states of the [FeFe]-hydrogenase from *Desulfovibrio desulfuricans*. *Journal of the American Chemical Society*, 128(51), 16808-16815.

[86] Dementin, S., Belle, V., Bertrand, P., Guigliarelli, B., Adryanczyk-Perrier, G., De Lacey, A., Fernandez, V., Rousset, M., & Léger, C. (2006). Changing the ligation of the distal [4Fe4S] cluster in NiFe hydrogenase impairs inter- and intramolecular electron transfers. *Journal of the American Chemical Society*, 128(15), 5209-5218.

[87] Krassen, H., Stripp, S., von, Abendroth. G., Ataka, K., Happe, T., & Heberle, J. (2009). Immobilization of the [FeFe]-hydrogenase CrHydA1 on a gold electrode: Design of a catalytic surface for the production of molecular hydrogen. *Journal of Biotechnology*, 142(1), 3-9.

[88] Zadvornyy, O., Lucon, J., Gerlach, R., Zorin, N., Douglas, T., Elgren, T., & Peters, J. (2012). Photo-induced H_2 production by [NiFe]-hydrogenase from *T. roseopersicina* covalently linked to a Ru(II) photosensitizer. *Journal of Inorganic Biochemistry*, 106(1), 151-155.

[89] Reisner, E., Powell, D., Cavazza, C., Fontecilla-Camps, J., & Armstrong, F. (2009). Visible Light-Driven H_2 Production by Hydrogenases Attached to Dye-Sensitized TiO_2 Nanoparticles. *J. Am. Chem. Soc.*, 131(51), 18457-18466.

[90] Morra, S., Valetti, F., Sadeghi, S., King, P., Meyer, T., & Gilardi, G. (2011). Direct electrochemistry of an [FeFe]-hydrogenase on a TiO_2 Electrode. *Chemical Communications*, 47(38), 10566-10568.

[91] Brown, K., Wilker, M., Boehm, M., Dukovic, G., & King, P. (2012). Characterization of Photochemical Processes for H_2 Production by CdS Nanorod-[FeFe] Hydrogenase Complexes. *Journal of the American Chemical Society*, 143(12), 5627-5636.

[92] Brown, K., Dayal, S., Ai, X., Rumbles, G., & King, P. (2010). Controlled Assembly of Hydrogenase-CdTe Nanocrystal Hybrids for Solar Hydrogen Production. *Journal of the American Chemical Society*, 132(28), 9672-9680.

[93] Mc Donald, T., Svedruzic, D., Kim-H, Y., Blackburn, J., Zhang, S., King, P., & Heben, M. (2007). Wiring-up hydrogenase with single-walled carbon nanotubes. *Nano Letters*, 7(11), 3528-3534.

[94] Blackburn, J., Svedruzic, D., Mc Donald, T., Kim-H, Y., King, P., & Heben, M. (2008). Raman spectroscoipy of charge transfer interaction between single wall carbon nanotubes and [FeFe] hydrogenase. *Dalton Transactions*, 5454-5461.

[95] Kihara, T., Liu-Y, X., Nakamura, C., Park-M, K., Yasuda-W, S., Qian-J, D., Kawasaki, K., Zorin, N., Yasuda, S., Hata, K., Wakayama, T., & Miyake, J. (2001). Direct electron transfer to hydrogenase for catalytic hydrogen production using a single-walled carbon nanotubes forest. *International Journal of Hydrogen Energy*, 36(13), 7523-7529.

[96] Noda, K., Zorin, N., Nakamura, C., Miyake, M., Gogotov, I., Asada, Y., Akutsu, H., & Miyake, J. (1998). Langmuir-Blodgett film of hydrogenase for electrochemical hydrogen production. *Thin Solid Films*, 327-329, 639-642.

[97] Elgren, T, Zadvorny, O, Brecht, E, Douglas, T, Zorin, N, Maroney, M, & Peters, J. (2005). Immobilization of active hydrogenases by encapsulation in polymeric porous gels. *Nano Letters*, 5(10), 2085-2087.

[98] Zadvorny, O., Barrows, A., Zorin, N., Peters, J., & Elgren, T. (2010). High level oh hydrogen production activity achieved for hydrogenase encapsulated in sol-gel material doped with carbon nanotubes. *Journal of Materials Chemistry*, 20-1065.

[99] Fontecave, M., & Artero, V. (2011). Bioinspired catalysis at the crossroads between biology and chemistry: A remarkable example of an electrocatalytic material mimicking hydrogenases. *Compte Rendu Chimie*, 14(4), 362-371.

[100] Dubois, M., & Dubois, D. (2009). The roles of the first and second coordination spheres in the design of molecular catalysts for H_2 production and oxidation. *Chemical Society Reviews*, 38(1), 62-72.

[101] Le Goff, A., Artero, V., Jousselme, B., Tran, P. D., Guillet, N., Metaye, R., Fihri, A., Palacin, S., & Fontecave, M. (2009). From Hydrogenases to Noble Metal-Free Catalytic Nanomaterials for H_2 Production and Uptake. *Science*, 326(5958), 1384-1387.

[102] Yahiro, A., Lee, S., & Kimble, D. (1964). Bioelectrochemistry I. Enzyme utilizing Biofuel cell studies. *Biochimica Biophysica Acta*, 88, 375-383.

[103] Franks, Ashley. E., & Nevin, Kelly. P. (2010). Microbial Fuel Cells. *A Current Review. Energies*, 3(5), 899-919.

[104] Zhou, M., Chi, M., Luo, J., He, H., & Jin, T. (2011). An overview of electrode materials in microbial fuel cells. *Journal of Power Sources*, 196(10), 4427-4435.

[105] Sokic-Lazic, D., & Minteer, S. (2008). Citric acid cycle biomimic on a carbon electrode. *Biosensors Bioelectronics*, 24(4), 939-944.

[106] Egorova, K., & Antranikian, G. (2005). Industrial relevance of thermophilic Archaea. *Current Opinion in Microbiology*, 8(6), 649-655.

[107] Halamkova, L., Halamek, J., Bocharova, V., Szczupak, A., Alfonta, L., & Katz, E. (2012). Implanted Biofuel Cell Operating in a Living Snail. *Journal of the American Chemical Society*, 134(11), 5040-5043.

[108] Cinquin, P., Gondran, C., Giroud, F., Mazabrard, S., Pellissier, A., Boucher, F., Alcaraz, J. P., Gorgy, K., Lenouvel, F., Mathé, S., Porcu, P., & Cosnier, S. (2010). A Glucose BioFuel Cell Implanted in Rats. *PLoS ONE*, 5, e10476.

[109] Falk, M., Andoralov, V., Blum, Z., Sotres, J., Suyatin, D., Ruzgas, T., Arnebrant, T., & Shleev, S. (2012). Biofuel cell as a power source for electronic contact lenses. *Biosensors Bioelectronics*, 37(1), 38-45.

[110] Zebda, A., Gondran, C., Le Goff, A., Holzinger, M., Cinquin, P., & Cosnier, S. (2011). Mediatorless high-power glucose biofuel cells based on compressed carbon nanotube-enzyme electrodes. *Nature Com.*, 2-370.

[111] Zhao, H., Zhou, H., Zhang, J., Zheng, W., & Zheng, Y. (2009). Carbon nanotube-hydroxyapatite nanocomposite: A novel platform for glucose/O_2 biofuel cell. *Biosensors Bioelectronics*, 25(2), 463-468.

[112] Liu, J., Zhang, X., Pang, H., Liu, B., Zou, Q., & Chen, J. (2012). High-performance bioanode based on the composite of CNTs-immobilized mediator and silk film-immobilized glucose oxidase for glucose/O_2 biofuel cells. *Biosensors Bioelectronics*, 31(1), 170-175.

[113] Gao, F., Yan, Y., Su, L., Wang, L., & Mao, L. (2007). An enzymatic glucose/O_2 biofuel cell: Preparation, characterization and performance in serum. *Electrochemistry Communications*, 9(5), 989-996.

[114] Li, X., Zhou, H., Yu, P., Su, L., Ohsaka, T., & Mao, L. (2008). A miniature glucose/O_2 biofuel cell with single-walled carbon nanotubes-modified carbon fiber microelectrodes as the substrate. *Electrochemistry Communications*, 10(6), 851-854.

[115] Nazaruk, E., Sadowska, K., Biernat, J., Rogalski, J., Ginalska, G., & Bilewicz, R. (2010). Enzymatic electrodes nanostructured with functionalized carbon nanotubes for biofuel cell applications. *Analytical and Bioanalytical Chemistry*, 398(4), 1651-1660.

[116] Lim, J., Malati, P., Bonet, F., & Dunn, B. (2007). Nanostructured sol-gel electrodes for biofuel cells. *Journal of the Electrochemical Society*, 154(2), A140-A145.

[117] Karaskiewicz, M., Nazaruk, E., Zelzchowska, K., Biernat, J., Rogalski, J., & Bilewicz, R. (2012). Fully enzymatic mediatorless fuel cell with efficient naphthylated carbon nanotube-laccase composite cathodes. *Electrochemistry Communications*, 20-124.

[118] Tasca, F., Gorton, L., Harreither, W., Haltrich, D., Ludwig, R., & Nöll, G. (2008). Highly efficient and versatile anodes for biofuel cells based on cellobiose dehydrogenase from *Myriococcum thermophilum*. *Journal of Physical Chemistry C*, 112(35), 13668-13673.

[119] Wang, Y., & Yao, Y. (2012). Direct electron transfer of glucose oxidase promoted by carbon nanotubes is without value in certain mediator-free applications. *Microchimica Acta*, 176(3-4), 271-277.

[120] Vincent, K., Parkin, A., & Armstrong, F. (2007). Investigating and exploiting the electrocatalytic properties of hydrogenases. *Chemical Reviews*, 107(10), 4366-4413.

[121] Liebgott, P. P., de Lacey, A., Burlat, B., Cournac, L., Richaud, P., Brugna, M., Fernandez, V., Guigliarelli, B., Rousset, M., Léger, C., & Dementin, S. (2011). Original Design of an Oxygen-Tolerant [NiFe] Hydrogenase: Major Effect of a Valine-to-Cysteine Mutation near the Active Site. *Journal of the American Chemical Society*, 133(4), 986-997.

[122] Cracknell, J. A., Vincent, K. A., Ludwig, M., Lenz, O., Friedrich, B., & Armstrong, F. A. (2008). Enzymatic Oxidation of H_2 in Atmospheric O_2: The Electrochemistry of Energy Generation from Trace H_2 by Aerobic Microorganisms. *Journal of the American Chemical Society*, 130(2), 424-425.

[123] Luo, X. J., Brugna, M., Infossi, P., Giudici-Orticoni, M. T., & Lojou, E. (2009). Immobilization of the hyperthermophilic hydrogenase from *Aquifex aeolicus* bacterium onto gold and carbon nanotube electrodes for efficient H_2 oxidation. *Journal of Biological Inorganic Chemistry*, 14(8), 1275-1288.

[124] Svedruzic, D., Blackburn, J., Tenent, R., Rocha, D. J., Vinzant, T., Heben, M., & King, P. (2011). High-performance hydrogen production and oxidation electrodes with hydrogenase supported on metallic single-wall carbon nanotubes networks. *Journal of the American Chemical Society*, 133(12), 4299-4306.

[125] Sun, Q., Zorin, N. A., Chen, D., Chen, M., Liu, X. T., Miyake, J., & Qian, J. D. (2010). Langmuir-Blodgett films of pyridyldithio-modified multiwalled carbon nanotubes as a support to immobilize hydrogenase. *Langmuir*, 26(12), 10259-10265.

[126] Hoeben, F. J. M., Heller, I., Albracht, S. P. J., Dekker, C., Lemay, S. G., & Heering, H. A. (2008). Polymyxin-coated Au and Carbon nanotubes electrodes for stable [NiFe]-hydrogenase film voltammetry. *Langmuir*, 24(11), 5925-5931.

[127] Baur, J., Le Goff, A., Dementin, S., Holzinger, M., Rousset, M., & Cosnier, S. (2011). Three-dimensional carbon nanotube-polypyrrole-[NiFe] hydrogenase electrodes for the efficient electrocatalytic oxidation of H_2. *International Journal of Hydrogen Energy*, 36(19), 12096-12101.

[128] Ciaccafava, A., Infossi, P., Ilbert, M., Guiral, M., Lecomte, S., Giudici-Orticoni, M. T., & Lojou, E. (2012). Electrochemistry, AFM, and PM-IRRA Spectroscopy of Immobilized Hydrogenase: Role of a Hydrophobic Helix in Enzyme Orientation for Efficient H_2 Oxidation. *Angewandte Chemie International Edition*, 51(4), 953-956.

[129] Volbeda, A., Amara, P., Darnault, C., Mouesca, J. M., Parkin, A., Roessler, M. M., Armstrong, F. A., & Fontecilla-Camps, J. C. (2012). X-ray crystallographic and computational studies of the O_2-tolerant [NiFe]-hydrogenase 1 from *Escherichia coli*. *Proceeding of the National Academic Sciences*, 10(14), 5305-5310.

[130] Shomura, Y., Yoon, S. K., Nishihara, H., & Higuchi, Y. (2011). Structural basis for a [4Fe-3S] cluster in the oxygen-tolerant membrane-bound [NiFe]-hydrogenase. Nature NIL_143., 479(7372), 253.

[131] Pandelia, M., Fourmond, V., Tron, P., Lojou, E., Bertrand, P., Léger, C., Giudici-Orticoni, M. T., & Lubitz, W. (2010). Membrane-Bound Hydrogenase I from the Hyperthermophilic Bacterium *Aquifex aeolicus:* Enzyme Activation, Redox Intermediates and Oxygen Tolerance. *Journal of the American Chemical Society*, 132(20), 6991-7004.

[132] Fritsch, J., Scheerer, P., Frielingsdorf, S., Kroschinsky, S., Friedrich, B., Lenz, O., & Spahn, C. M. T. (2011). The crystal structure of an oxygen-tolerant hydrogenase uncovers a novel iron-sulphur centre. Nature NIL_134., 479(7372), 249.

[133] Krishnan, S., & Armstrong, F. A. (2012). Order-of-magnitude enhancement of an enzymatic hydrogen-air fuel cell based on pyrenyl carbon nanostructures. *Chemical Science*, 3(4), 1015-1023.

[134] Vincent, K., Cracknell, J., Clark, J., Ludwig, M., Lenz, O., Friedrich, B., & Armstrong, F. (2006). Electricity from low-level H_2 in still air- an ultimate test for an oxygen tolerant hydrogenase. *Chemistry Communications*, 5033-5035.

[135] Wait, A., Parkin, A., Morley, G., dos, Santos. L., & Armstrong, F. (2010). Characteristics of enzyme-based hydrogen fuel cells using an oxygen-tolerant hydrogenase as the anodic catalyst. *Journal of Physical Chemistry C*, 114(27), 12003-12009.

[136] Ciaccafava, A., de Poulpiquet, A., Techer, V., Giudici-Orticoni, M. T., Tingry, S., Innocent, C., & Lojou, E. (2012). An innovative powerful and mediatorless H_2/O_2 biofuel cell based on an outstanding bioanode. *Electrochemistry Communications*, 23, 25-28.

[137] Infossi, P., Lojou, E., Chauvin, J. P., Herbette, G., Brugna, M., & Giudici-Orticoni, M. T. (2010). *Aquifex aeolicus* membrane hydrogenase for hydrogen bioxidation: role of lipids and physiological partners in enzyme stability and activity. *International Journal of Hydrogen Energy*, 35(19), 10778-10789.

[138] Reuillard, B., Le Goff, A., Agnès, A., Zebda, A., Holzinger, M., & Cosnier, S. (2012). Direct electron transfer between tyrosinase and multi-walled carbon nanotubes for bioelectrocatalytic oxygen reduction. *Electrochemistry Communications*, 20, 19-22.

[139] Durand, F., Kjaergaard, C., Suraniti, E., Gounel, S., Hadt, R., & Solomon, E. (2012). Mano N Bilirubin oxidase from *Bacillus pumilus*: A promising enzyme for the elaboration of efficient cathodes in biofuel cells. *Biosensors Bioelectronics*, 35(1), 140-146.

Chapter 7

Interconnecting Carbon Nanotubes for a Sustainable Economy

Steve F. A. Acquah, Darryl N. Ventura,
Samuel E. Rustan and Harold W. Kroto

Additional information is available at the end of the chapter

1. Introduction

Concerns about depleting natural resources have been circulating for decades with alarming predictions that have turned out to be less than accurate. What has become clear, however, is the need for a decrease in the utility of a fossil based economy and a focus on a more sustainable one. This chapter reviews some of the recent progress made in the use of interconnected carbon nanotubes (CNTs) in the hydrogen, photovoltaics and thermoelectric alternative energy based economies.

The move towards a hydrogen economy is a concept that has gained traction over the last 5 years with advances in hydrogen fuel cells that are economically viable. It is envisaged that the automotive industry will begin to implement measures for the development of vehicles with hydrogen fuel cells as the economy begins to recover. However, such a move will also require a substantial investment in the infrastructure to support these vehicles. Key to the development of such technologies is the need to continuously improve the efficiency, while monitoring the safety. CNTs have been used as frameworks for a number of key areas in the hydrogen economy [1]. The most notable area is that of fuel cell integration, where the tubes are mixed with platinum or palladium to aid in the process of catalysis.

CNTs with palladium attached to their surface have also been used for the construction of hydrogen sensors, expanding the research field from the consumption to the detection of hydrogen. The recent advances in cross-linked CNT papers are stimulating the development of new materials, such as flexible palladium embedded CNT sensors [2] (Fig. 1.). This section of the chapter will explore some of the latest results from the use of interconnected CNTs in hydrogen fuel cells and sensor development.

Figure 1. A cross-linked CNT paper with embedded Pd nanoparticles that can be used to construct a hydrogen sensor.

The field of photovoltaics is regarded as a major contributor to a sustainable economy. However; purveyors of large scale solar panels have been experiencing a degree of volatility in the market due in part to the decreasing price of the technology, increased competition and a dependence on government subsidies. At the opposite end of the scale, there is a surge in small solar powered gadgets such as pocket LED torches and mobile device chargers, which adorn many airport convenience outlets. The demand for pocket sized solar powered devices is helping to stimulate research into making the energy conversion process more efficient. There were three major advances in photovoltaics, the development of photovoltaic devices from crystalline silicon, which dominate the commercial market, cadmium telluride (CdTe) and dye sensitized solar cells (DSSCs). CNTs are currently being investigated as a way to enhance electron transfer and replace the standard platinum based counter electrodes, especially with DSSCs. CNT thin films and mats are currently being tested as components of these photovoltaic devices. This section of the chapter will explore how the CNTs have been used to enhance dye-sensitized [3], CdTe [4] and silicon [5] based solar cells, and address some the concerns about the race to produce novel photovoltaic devices and the toxic warnings from the past that may ultimately define the balance between safety and efficiency.

The last section of this chapter will focus on the development of CNT based thermoelectric devices which may bridge the gap between conventional and sustainable economies. Energy loss in the form of heat is clearly an important concept to address, and capturing the heat from combustion engines is one avenue being pursued by research. Around 75% of the energy produced from fuel with internal combustion engines is lost to the environment, so it may be possible to recapture some of this energy using a thermoelectric device between the engine coolant system to the exhaust manifold [6]. However, problems have been encountered with low efficiency so CNTs have been investigated as a suitable component of thermoelectric devices due to a number of characteristics, such as their low dimensional structure, their electrical conductivity, and their axial thermal conductivity [7, 8].

2. The Hydrogen Economy

Many nations are looking into alternative sources of energy to address issues of environmental responsibility and energy independence. Some of these energy sources include solar power, wind energy, natural gas, and hydrogen. As society explores hydrogen as an alternative energy source, the question is how effective can CNTs be in helping to solve some of the problems in the structure, function and safety of this emerging industry?

2.1. Fuel Cells & Hydrogen Storage

In the simplest case, a hydrogen fuel cell is comprised of a permeable membrane placed between an anode and a cathode. There are various types of fuel cells: polymer electrolyte membrane, direct methanol, alkaline, phosphoric acid, molten carbonate, and solid oxide. Hydrogen fuel cells fall under the polymer electrolyte membrane fuel cell (PEMFC) category and are sometimes also referred to as a proton exchange membrane fuel cell. In a typical PEMFC, the permeable membrane consists of a proton-conductive polymer such as perfluorosulphonic acid, also known commercially as Nafion. The fuel cell works by using a catalyst to oxidize hydrogen at the anode, converting it into a positively charged proton and a negatively charged electron. The electrons travel through a wire creating an electrical current to power a device while the protons travel through the permeable membrane to the cathode. At the cathode, the protons recombine with the electrons and react with oxygen to form water which is eventually drained from the system.

Despite recent advances in research, there are still a few obstacles that need to be overcome in order for fuel cells to become mainstream technology. In order to integrate with existing technologies, fuel cells need to become considerably cheaper. Currently, they are expensive to construct, mainly due to the use of platinum catalysts. According to the United States Department of Energy, the cost per kilowatt would need to decrease in order for fuel cells to be competitive and economically viable. In order to compete commercially with the combustion engine, it is estimated that the fuel cell cost would need to be cut to approximately $25–$35/kW. Another aspect of fuel cells that needs improvement is the operational lifetime. The permeable membrane is made of a synthetic polymer which is susceptible to chemical degradation. Reliability in automotive applications, can be defined by the lifetime of a car engine, approximately 150,000 miles, so research has focused both on improving the efficiency of the catalytic process and the durability of the components. CNTs have been proposed as a substitute to the carbon powder currently used in PEMFCs. (Fig. 2.) CNTs have excellent conductive properties, a low mass density, and robust physical properties making them an ideal and durable material for fuel cell electrodes. Furthermore, nanotubes assembled into such macrostructures have a high surface area making them a suitable substrate for Pt catalysts and hydrogen adsorption [9].

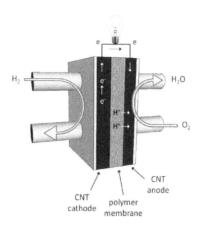

Figure 2. Schematic of a CNT composite hydrogen fuel cell.

In 2003, researchers from the University of California, Riverside explored the use of MWCNTs as a carbon support for platinum catalysts in an attempt to maximize Pt interfacing between all the components in a fuel cell. The problem in conventional fuel cells is that the addition of the polymer tends to isolate the carbon particles reducing electron transport, resulting in the requirement of additional Pt particles to increase the power output. To resolve this issue and improve conductivity, Wang et al. grew nanotubes directly on carbon paper and electrodeposited Pt particles onto the CNTs [10]. Although their experiments produced promising results, their CNT based fuel cell still had a lower performance compared to conventional PEMFCs. Despite this low performance, this proof of concept was important to other researchers using CNTs in fuel cells. The following year in 2004, Girishkumar et al. investigated ways to improve the electrodes in direct methanol fuel cells (DMFCs) [11]. Their team developed a way to synthesize SWCNT thin films onto optically transparent electrodes using electrophoretic deposition techniques. It was determined that there was an improvement in catalytic activity mainly due to a larger surface area provided by the CNTs. This high surface area and porosity maximizes interactions between the fuel, electrode, and catalyst interface thereby enhancing Pt utilization and potentially reducing fuel cell manufacturing costs. Li et al. (2006) also explored the use of CNTs in PEMFCs. They developed a facile and cost-effective method for the synthesis of an aligned Pt/CNT film [12]. They were interested in producing oriented CNT films due to enhanced conductivity. It was also suggested that there would be higher gas permeability and better water removal with aligned nanotubes. The aligned CNTs did show an improvement in Pt utilization as 60% of the metal particles were being used during catalysis [11].

Using covalently cross-linked CNTs is another promising avenue for fuel cell electrodes [13]. Our work at Florida State University focused on the covalent cross-linking of multi-walled carbon nanotubes via a Michael addition reaction mechanism to form thin, flexible mats

[14]. We then explored an alternative cross-linking system to avoid the use of thiols and embedded palladium nanocrystals into the cross-linked network [2].

Research into hydrogen storage with interconnected CNT networks started by looking into SWCNTs using a procedure called temperature programmed desorption. Experiments on MWCNTs followed with work focusing on metal doped tubes. However, problems began to arise when increasing values of CNT storage capacities, up to 21 wt%, were reported. A detailed review of the findings can be found by Yunjin Yao and serves as an interesting footnote towards the role of CNTs and the need for a better understanding of their chemistry in materials [15]. In summary, the main concerns were that elevated hydrogen storage percentages may have be due to a number of factors including the insufficient characterization of CNT composites due to the presence of SWCNTs, DWCNTs and MWCNTs with a variety of open and closed ended tubes. Contamination of the CNTs during the process of ultrasonic probe treatments was a concern, because in one example the value for SWCNTs were reported to have a hydrogen storage capacity of around 4.5% at 30 kPa and 70 K, but the ultrasonic probe was made from a titanium alloy that was known to act as a hydrogen storage material.

2.2. Water Splitting

The research field based on water splitting has, not surprisingly, found a niche in the development of the hydrogen economy due to the clean production of hydrogen and oxygen. However this integration has a far more significant impact when combined with hydrogen fuel cells. The waste product of hydrogen fuel cells is water, and it is formed during the reaction with oxygen, so the water could fuel the process of splitting and this in turn can fuel hydrogen cells.

CNTs have been used to enhance the water splitting performance of titania photocatalysts [16] but an alternative use for CNTs has been found in membranes. Nafion, a sulfonated tetrafluoroethylene based fluoropolymer-copolymer, is a membrane that has had commercial success in the fuel cell industry. Research groups are looking into enhancing the properties of the film with the addition of CNTs. Nafion/CNT composites with low concentrations of CNTs have been shown to have an effect on solvent permeation and mechanical stability. At high concentrations of CNTs the membranes have the ability to separate proton and electron conduction pathways in the membrane. Using this concept, many applications can be envisaged for these membranes with one example being that of using sunlight to produce hydrogen from water splitting. Current research has focused on the measurements of the electron and proton transport characteristics of Nafion and MWCNT composite films.[17] These films can be assembled by the addition of Nafion solution to MWCNTs, followed by the dispersion of the MWCNTs in an ultrasonic bath. Various concentrations of MWCNTs were investigated to a maximum of 5% MWCNTs by dry weight of Nafion. After the addition of isopropyl alcohol, to further aid the dispersion of the MWCNTs, the slurry was poured into petri-dishes and left to undergo solvent evaporation for 3 hours. The dishes with various

CNT concentrations were placed in an oven set to 40 °C for a few hours before being washed with deionized water and removed from the petri-dishes.

To test the membrane, an artificial leaf system was constructed. (Fig. 3.). The membrane separated the anode, which was exposed to sunlight where water droplets were present, and the cathode.

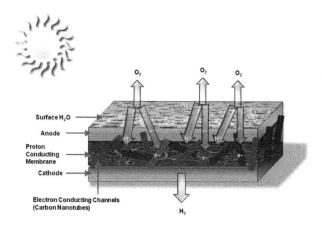

Figure 3. Schematic of the water-splitting device. The anode contained a chromophore and an oxygen evolving complex. The cathode contained a proton reducing catalyst. Image adapted from V. Ijeri *et al.* (2010).

The results highlighted a few points. Firstly pure Nafion exhibited insulating behavior and with increasing MWCNT percentage, a non linear behavior is observed with I–V curves, which is an indication of non-ohmic conductivity. The membranes were tested in both wet (1% H_2SO_4) and dry conditions to evaluate the electron conductivity. Before and after wetting the conductivity values increase with increasing filler content, but again without a linear relationship, which meant a critical concentration at which the membrane changed from insulating to conducting/semiconducting had to be established. This was done by looking at the values higher than 10^{-1} mS/cm which were obtained when MWCNTs > 3%. The next task was to investigate proton conductivity, and with standard conditions, this was generally low. However, with an increasing MWCNT percentage there was a subsequent increase. Although the effects of MWCNTs on proton conductivity is still not fully understood, most researchers will fall back on the semi-empirical quantum mechanical calculations too at least provide an insight into the possible conduction pathways.

When the membranes were subjected to 1% H_2SO_4 they did show an increase in proton conductivity, which was due to the various proton transfer mechanisms. The hydrogen bonding of the –SO_3 groups with an H_3O^+ ion and water molecules results in a change in the side chains of Nafion. It was difficult to determine the contribution of MWCNTs in the process of electron transfer because of the amount of water molecules. However membranes with no MWCNTs demonstrated the best proton conductivity, while the others have slightly lower conductivities. The answer could be as simple as a decrease in the amount of Nafion. Either way, this study has shown great potential for the integration of CNTs for membrane applications.

2.3. CNT Hydrogen Sensors

Another application of great interest in the field of CNTs is hydrogen sensing. Advancements in the development of fuel cell design and technology means that a variety of sensors would be required to maintain a safe operational environment. CNTs are an ideal material for components of sensors due to their durability, and electronic properties.

One of the first breakthroughs in CNT sensor technology occurred in 2001 when Kong et al. constructed hydrogen sensors by decorating SWCNTs with Pd nanoparticles [18]. Their H_2 sensor exhibited significant changes in conductivity when exposed to small amounts of H_2 and was able to operate at room temperature. Kong et al. were able to achieve this by depositing Pd particles on CVD grown SWCNTs via electron beam evaporation methods. When they placed this in a hydrogen atmosphere, a decrease in the CNT conductivity was observed. It has been proposed that this lower work function promotes electron transport from the Pd NPs into the CNTs resulting in a decreased amount of hole-carriers and conductivity. The reaction is also reversible. Under a hydrogen atmosphere, Pd reacts with H_2 to become palladium hydride. The dissolved hydrogen in Pd metal combines with oxygen in air and results in H_2O, recovering the electrical characteristics of the sensor. Kong et al. reported that their detector had a limit at 400 ppm, a response time of 5-10 s, and a recovery time of approximately 400 s [18].

One design principle of CNT composites that has defined the nature of efficiency is that of aligned CNTs. From aligned thin films of Buckypaper to forests of vertically grown CNTs on substrates, control over the direction of individual tubes and connected bundles is essential for unlocking the full potential of the tubes. An investigation was made into the development of aligned CNT sensors using a method involving nanoplating and firing to produce cracks in a CNT composite film, exposing horizontally aligned carbon nanotubes (HACNTs) [19]. This research used arc produced MWCNTs as the basis for the composite film, which was rather enlightening in a field geared towards chemical vapor deposition (CVD) produced tubes. Research with arc produced MWCNTs has almost become a relic of the early years of CNT research. They were made by using a 150 mm long graphite rod for the anode and a graphite disc on a copper block for the cathode. After purification steps the CNTs were acid-oxidized using the standard 3:1 ratio of nitric acid (HNO_3) to sulfuric acid (H_2SO_4) and washed several times in DI water before being dried in air at 120 °C. The acid treatment was required to increase the interfacial adhesion between the CNTs and metals. To produce the sensor, a sample of the purified CNTs was dispersed in DI water with a polyvinylpyrrolidone surfactant (PVP K30), which produced a CNT suspension. The CNT/Ni composite was produced by the addition of nickel sulfate solution containing sodium phosphinate, maleic acid disodium salt hydrate, citric acid monohydrate, lead(II) acetate trihydrate and sodium acetate trihydrate. The composite film was produced on a glass substrate by the immersion of the glass, with palladium particles on the surface, into the CNT/Ni solution for 60 seconds before drying the substrate at 100 °C to induce cracks in the film (Fig. 4.) exposing horizontally aligned CNTs. 18 Finger platinum electrodes were then deposited by DC sputtering to complete the sensor.

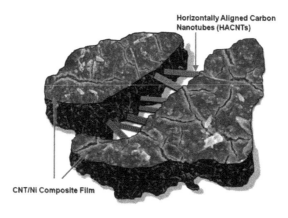

Figure 4. Schematic of the cracked composite film exposing horizontally aligned CNTs.

These results are described in two papers and although the idea of horizontal alignment is important, it is difficult to accurately quantify the results of the papers since in both cases there is an abundance of nanoparticle palladium in both the CNT/Ni system (Pd deposited on the glass) [19] and the Pd/CNT/Ni (Pd deposited on the CNT/Ni film) system [20]. Fig. 5. shows the process of assembly for the sensors, which use a similar procedure in both of the research papers.

The HACNT-based sensors were also shown to have a sensitivity response to carbon dioxide, methane and ethene with a gas concentration of 200 ppm, with the highest sensitivity for H_2. One of the points raised in this research, that was fundamental to the mechanism of sensing, was the role of atomized hydrogen. These atoms, produced by the metal particles, migrated to the sidewalls and the defects of CNTs, diffusing into the lattice of nanoparticles. It was stated that a dipole layer formed at that interface and affected the charge-carrier concentration, and the hydrogen atoms donated their electrons to the CNTs, which resulted in a decrease in conductivity.

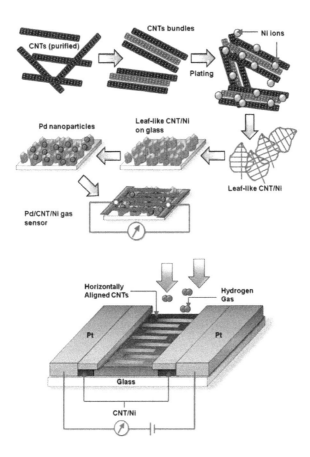

Figure 5. Schematic of the steps involved in the construction of hydrogen sensors on glass substrates with the use of the Pd nanoparticle functionalized CNT/Ni composite film. Image adapted from Lin *et al.* (2012). Schematic illustration of a HACNT-based gas sensor on glass substrate. Image adapted from B-R Huang *et al.* (2012).

In another example, a hydrogen sensor was constructed using SWCNTs and chitosan (CHIT).[21] The CHIT which covered the SWCNTs was able to filter out polar molecules and allow hydrogen to flow to the SWCNTs. The CHIT conjugate which is porous is insulating by nature, but can be made water soluble in an acidic environment which is then useful for making a film. Additional benefits can be found in the many functional hydroxyl (–OH) and amino (–NH$_2$) groups that react with analytes, so the effect Of a CHIT conjugate with SWCNTs for the development of a hydrogen sensor was investigated. The CHIT film was prepared by making a 2 wt% solution dissolving CHIT in a 5% acetic acid solution. This was used to coat a glass substrate or SWCNTs depending on the sensor preparation and followed by the removal of solvent to form the films. To evaluate the sensor performance three different types were made (Fig. 6.). The Type I sensor was assembled simply by depositing SWCNTs onto the glass substrate with Pt electrodes placed by sputter deposition. The Type II sensor was assembled by casting the glass slide with a film of CHIT before being placed into an arc-discharge chamber to deposit SWCNTs. The Pt electrodes were added in a similar method. The Type III sensor was assembled using the initial preparation for a Type I sensor followed by CHIT film coating and Pt electrode deposition. There were slight differences in the interaction of the CHIT film with the SWCNTs. In the Type II sensor, there was some mixing of the CNTs with CHIT but only at the interface. With the Type III sensor, the CNTs were immersed in the CHIT matrix.

Resistance measurements of the films were made between the electrodes, and the values were around 100 Ω for Type I and II films and around 10^6 Ω for the Type III film. The high resistance could be accounted for by the contact of the electrode with chitosan, although it was noted by the authors that ohmic contacts were present.

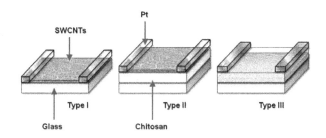

Figure 6. Diagram of the 3 types of sensors. Image adapted from Li *et al.* (2010).

The response of the sensors was measured at room temperature and the results showed 15, 33, and 520% for Type I, Type II, and Type III sensors, respectively. One interesting point made by the authors was that although the Pd decoration of SWCNTs is typically used to enhance hydrogen sensing, the response can be less than the effect of chitosan at 4% H_2 gas. This research provided an important step towards the use of CNTs in sensors without the requirement of Pd.

In summary, the use of CNTs in the hydrogen economy has highlighted some interesting points. Is the race to develop more efficient hydrogen powered devices really producing a sustainable economy? And has the focus on reducing the utility of some of the rare raw materials been lost? It is well known that platinum and palladium are extremely important to the fuel cell and sensor industries, with CNTs enhancing their properties, but an increase in alternative energy devices based on these metals, whatever the concentration, may cause issues of sustainability in the future.

3. Photovoltaics

The research field of photovoltaics has certainly become a hot topic over the last few years with a lot of attention based on increasing the efficiency of dye sensitized solar cells (DSSCs) in the hope that they will one day be as prevalent as the silicon based alternative. CNTs are an important addition to the field of photovoltaics with the focus on the nanotubes acting as p-type materials or enhancing/replacing the counter electrodes.

3.1. Dye Sensitized Solar Cells

If there were an enclave for truly beautiful chemistry, then the research behind dye sensitized solar cells (DSSCs) would clearly be the centerpiece. The chemistry behind the operation of these devices is inspiring a generation of researchers to address the concerns of renewable energy with a different approach to the well established silicon based solar cells. Generally, the DSSCs are comprised of an anode, electrolyte and cathode. The anode is usually assembled from nano-crystalline titania particles (TiO_2) and a dye attached to the particles. The cathode, also known as the counter electrode (CE), is where the catalysis must occur and typically contains platinum. The iodide electrolyte facilitates the iodide/triiodide redox couple where after the excitation of the dye and loss of an electron, it regains one from iodide, oxidizing it to triiodide. The best reported efficiency for DSSCs is 11.4% as documented by the National Institute for Material Science (NIMS).

CNTs have been used as a potential replacement for the platinum based CE. In a study by Jo et al. (2012), interconnected ordered mesoporous carbon–carbon nanotube nanocomposites were used to demonstrate Pt-like CE behavior in a dye-sensitized solar cell [22]. CNT fibers have been used as a conductive material to support the dye-impregnated TiO_2 particles. The CNTs were first spun from an array synthesized by chemical vapor deposition and resulted in highly aligned macroscopic fibers [23]. The research was novel in the application of these fibers as both the working electrode and the counter electrode.

The CNT/TiO_2 composite fiber was produced by submersing the pure CNT fiber in a TiO_2 colloid solution which was followed by sintering at 500 °C for 60 min. The thickness of TiO_2 layer was determined to be between 4 and 30 μm, depending on the submersion time. The dye used for the cell was cis-diisothiocyanato-bis (2,2′-bipyridyl-4,4′-dicarboxylato) ruthenium(II) bis (tetrabutylammonium) which is better known as N719. For DSSCs with a metal CE the I^-/I_3^- couple does eventually cause corrosion, but the CNT fibers exhibit a high stability and are relatively cheap. Fig. 7. shows the schematic of the working device with the two fibers in an electrolytic solution.

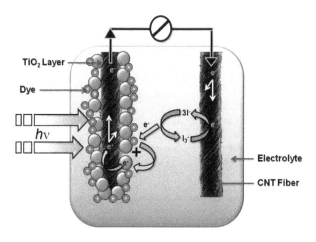

Figure 7. Schematic illustration of a wire-shaped DSSC made from two CNT fibers. Figure adapted from Chen *et al.* (2012).

The mechanical properties of the fiber are quite good with tensile strength measurements that exceed 700 MPa. The enhanced electrical conductivity also ranges from 100 to 1000 S/cm. The fiber-shaped DSSC demonstrated an efficiency of 2.94% which was a significant accomplishment. The fibrous nature of the material would make large-scale composites easy to fabricate. One of the more exciting applications is that of woven fabrics that may be used for the development of smart textiles for consumers, or extended use for space based electronics.

3.2. Quantum Dot Solar Cells

Cadmium telluride (CdTe) has been shown to be a promising low-cost component photovoltaic material, however the incorporation of quantum dot (QD) based technologies will likely raise fears about the toxicity of cadmium and cadmium based compounds. Significant progress has been made during the past several years with the highest efficiency reported for CdTe based photovoltaic devices at 17.3% produced by the company First Solar.

Although research is shifting towards CdTe/graphene composites [24], there is still room for CNT based devices. SWCNT/polyelectrolyte/QD nanohybrids have been produced that take advantage of the negatively charged thioglycolic acid capped CdTe QDs and SWCNTs coated with a positively charged polyelectrolyte facilitating electrostatic interactions [25]. In this work, SWCNTs coated with a positively charged polyelectrolyte showed typical transitions and emission attributes in the visible and near-infrared spectrum. The application of steady state absorption spectra was useful in outlining the superimposition of QD and SWCNT characteristics. The results of the study also confirmed charge transfer between SWCNTs and QDs, underlined by femtosecond transient absorption spectroscopy. Microscopic studies suggested that statically formed SWCNT/polyelectrolyte/QD nanohybrids with individually immobilized QDs were generated. It is clear that this study focuses on the importance of the interactions between the components of the nanohybrids and creates a pathway for looking at the development of the layer-by-layer coating of SWNTs and recruitment of photoactive particles for photovoltaic applications.

3.3. Silicon Based Solar Cells

With the exception of multi-junction cells and gallium arsenide (GaAs) based devices, crystalline silicon based cells are still the best choice with efficiencies at 20.4% for multicrystalline structures to 27.6% for single crystal based cells. However, there is clearly room for improvement as the increase in efficiency has generally reached a plateau over the last few years. What may be required is a different approach to the design and chemistry of these photovoltaic devices. CNTs have again been applied on the strength of their p-type conduction. In one recent example, polyaniline (PANI) and CNTs were used to construct heterojunction diode devices on n-Type silicon [26]. If was found that both PANI and SWCNTs could act as photovoltaic materials in a bilayer configuration with n-type Silicon: n-Si/PANI and n-Si/SWCNT. Four devices were tested (Fig. 8.) and it was determined that the short circuit current density increased from 4.91 mA/cm^2 for n-Si/PANI (Fig. 8a) to 12.41 mA/cm^2 n-Si/PANI/SWCNT (Fig. 8c). The n-Si/SWCNT/PANI device (Fig. 8d) and its control n-Si/SWCNT (Fig. 8b) exhibited a decrease in the short-circuit current density.

PANI was synthesized using the MacDiarmid method [27] before being spin-coated at 600 rpm to form a film. The SWCNTs were dispersed in DMF by sonication over a period of 12 h in 3 hour intervals, with the any solids removed by centrifugation. The supernatant was then removed and sonicated for an additional 6 hours before being used to make the devices. The devices were assembled by spraying SWCNTs using an airbrush deposition technique at 150 °C. It was found that the characteristics of the devices were affected by their design structure with better hole transport from PANI to SWCNTs and less efficient transport of holes from PANI to SWCNTs in the multilayer devices.

Other examples of CNT-Silicon hybrid photovoltaic devices include the investigation of the optimal thickness of SWCNT films on n-type silicon in order to maximize photovoltaic conversion [28] giving percentage efficiencies between 0.4 and 2.4%, and the effect of the number of walls of MWCNTs on the photon to electron conversion [29].

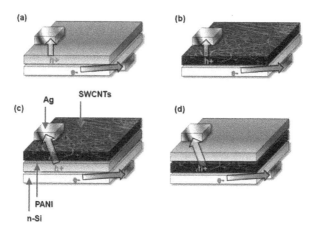

Figure 8. Schematics for (a) n-Si/PANI, (b) n-Si/SWCNTs, (c) n-Si/PANI/SWCNTs, and (d) n-Si/SWCNT/PANI devices. Image adapted from Bourdo *et al.* (2012).

In summary, photovoltaics have been shown to be very popular within the scientific field and the commercial market. Consumer electronics have been marketed with solar power chargers as a way to promote sustainability and environmental responsibility. The research into ruthenium based DSSCs is very popular but again there are concerns about the use of ruthenium for a sustainable economy. Fortunately, there are many photosensitive dyes that don't contain ruthenium which are currently being explored, but it is clear that the integration of interconnected CNTs can play an important role in the development of novel photovoltaic devices.

4. Thermoelectrics

In 1821 Thomas Johann Seebeck made the first discovery in the series of thermoelectric effects. The Seebeck effect described the electromotive force (emf) produced by heating the junction between two different metals. In essence, the kinetic energy of the electrons in the warmer part of a metal would facilitate the transfer of the electrons to the cooler metal faster than electron transfer from the cooler to the warmer metal, essentially creating an electronic potential where the cooler metal obtains a net negative charge. Harnessing the heat lost from a system and converting it to electricity will help to reduce the strain on electricity providers, but the difficulties surrounding the efficiency of the conversion process need to be addressed.

4.1. Thermoelectric Fabrics

One of the more futuristic ideas is that of wearable electronics, and this has been envisaged for many in the field of photovoltaics, but an Interesting alternative can be found in the field of thermoelectrics. Recent advancements in research have shown that composite films of MWCNT and polyvinylidene fluoride (PVDF) assembled in a layered structure can be designed to have the effect of felt-like fabric.[30] A thermoelectric voltage can be generated by these fabrics as a result of the individual layers increasing the amount of power produced. More importantly, these fabrics would be more economical to produce clearing the way for a new generation of energy harvesting devices that could power portable electronics. Fig. 9. shows a schematic of a fabric with every alternate conduction layer made with p-type CNTs (B) followed by n-type CNTs (D). The insulating layers allow for alternating p/n junctions when all the layers are stacked, pressed and heated to melt the polymer. It was noted that layers A–D could be repeated to reach a desired number of conduction layers N, and when the film is exposed to a change in temperature ($\Delta T = T_h - T_c$), the charge carriers which can be holes (h) or electrons (e) migrate from T_h to T_c generating a thermoelectric current I.

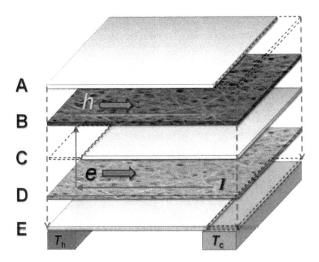

Figure 9. A layered arrangement for the multilayered fabric. The CNT/PVDF conduction layers (B,D) are alternated between the PVDF insulation layers (A,C,E). Figure adapted from Hewitt et al. (2012).

When more power is required, ΔT would have to be increased. Subsequently, if the heat source were sufficiently large enough, the number of conduction layers could be increased. This would be a huge benefit for manufacturing industries that use high temperature equipment. In terms of energy output, a fabric composed of 300 layers with a $\Delta T = 100$ K, may produce up to 5 μW. This is certainly a promising material that could potentially be integrated into many thermal systems and help with waste heat recovery.

4.2. Micro-Thermal Electrics

The addition of CNTs to microelectrical mechanical systems (MEMS) typically proceeds by either a bottom-up approach which focuses on the deposition of catalytic nanoparticles to control the location of CNT growth or a top-down which concerns the manipulation of the CNTs to the correct position. A top-down method was use to make a CNT thin film on a microelectrical mechanical system which was then characterized in terms of the thermoelectric coefficients of the aligned SWCNTs [8]. Using the process of 'super-growth' which incorporates water-assisted chemical vapor deposition, a CNT film was made and patterned by electron beam lithography into the required dimensions. By patterning a formed array of gold–SWCNT thermocouples it was found that under standard room temperature the Seebeck coefficient of the aligned SWCNT film was between 18 and 20 μV C^{-1}. The Seebeck effect of the SWCNT film was documented using thermocouples made of gold–SWCNT (Fig. 10.). Electrodes, a hot end and cold end temperature sensor, and a heater were produced by photolithography, and with a gold lift-off process on top of a silicon substrate that was covered by an insulating layer of Si$_3$N$_4$. The SWCNT film was then constructed on the gold surface using the process of top-down assembly.

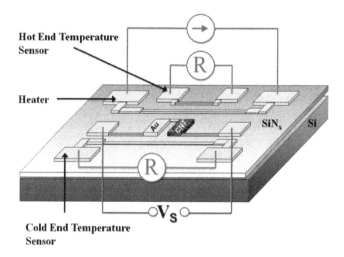

Figure 10. Schematic of a device for measuring the Seebeck effect in a CNT film. Figure Adapted from Dau *et al.* (2010).

When the device was used, an output voltage of 54 μV was recorded with a temperature difference of 3.07 °C. This gave a Seebeck voltage of 19.38 μV K^{-1} which on average remained constant. Aligned CNT bundles may have smaller Seebeck coefficients (thermoelectric sensitivity) than randomly oriented CNTs. The authors suggested that the difference may be a result of the contribution of inter-tube barriers, relative to ΔT, although more work is required to fully understand the effect of CNT films for the integration of them into thermoelectric devices.

5. Conclusion

CNTs have seldom been just another material for novel composites, but their true potential has yet to be transferred from the nano- to macro-scale. More than a two decades after their discovery, their influence has reached almost every aspect of scientific research from engineering to medicine. Faced with concerns about sustainably and climate change, the use of CNTs have helped to transform our approach to renewable energy. Advances in hydrogen fuel cells with CNT composite electrodes or membranes are helping to reduce and eliminate the need for rare and expensive catalysts. Safety is also another issue for the hydrogen based economy. Many different types of sensors will be required to promote a safe operational environment especially when the ignition concentration of hydrogen can be as low as 4%. The same technology that is used in the catalysis process in hydrogen fuel cells can be used to create hydrogen sensors, and work with interconnected CNTs has provided sensitivity values that contend with conventional sensors.

The role of interconnected CNTs in the photovoltaic research field is popular because of the potential to make novel hybrid solar cells, whilst increasing the overall efficiency of the device. While the early results look promising, there are still some difficult questions to address, like how does the presence of defects on the CNT surface affect the chemistry and ultimately the efficiency of a DSSC?

The integration of CNTs into thermoelectric devices currently does not have the same level of development as the other alternative energy resources, possibly because the field is more geared towards cost saving on an industrial scale and the development of component systems for vehicles rather than consumer gadgets or devices, but research into waste heat recovery is substantial. It is likely that thermoelectric devices will conform more to a silent revolution with an uptake in industries that work with high temperature equipment looking at converting some of the heat produced back to electricity. However, the research into thermoelectric fabrics has shown the potential for consumer products that may find a market in the future.

In summary, we are beginning to see a shift towards alternative fuel sources, with a focus on hybrid technologies like those found in the automotive industries, but we need to address the impact of our current economy as we transition to a more sustainable one.

Author details

Steve F. A. Acquah[*], Darryl N. Ventura, Samuel E. Rustan and Harold W. Kroto

*Address all correspondence to: acquah51@hotmail.com

Florida State University, United States

References

[1] Dillon, A. C., Jones, K. M., Bekkedahl, T. A., Kiang, C. H., Bethune, D. S., & Heben, M. J. (1997). Storage of hydrogen in single-walled carbon nanotubes. *Nature.*, 386(6623), 377-379.

[2] Ventura, D. N., Li, S., Baker, C. A., Breshike, C. J., Spann, A. L., Strouse, G. F., Kroto, H. W., & Acquah, S. F. A. (2012). A flexible cross-linked multi-walled carbon nanotube paper for sensing hydrogen. *Carbon.*, 50(7), 2672-2674.

[3] Velten, J., Mozer, A. J., Li, D., Officer, D., Wallace, G., Baughman, R., & Zakhidov, A. (2012). Carbon nanotube/graphene nanocomposite as efficient counter electrodes in dye-sensitized solar cells. *Nanotechnology.*, 23(8), 6.

[4] Barnes, T. M., Wu, X., Zhou, J., Duda, A., van de Lagemaat, J., Coutts, T. J., Weeks, C. L., Britz, D. A., & Glatkowski, P. (2007). Single-wall carbon nanotube networks as a transparent back contact in CdTe solar cells. *Applied Physics Letters*, 90(24).

[5] Jia, Y., Cao, A. Y., Bai, X., Li, Z, Zhang, L. H., Guo, N., Wei, J. Q., Wang, K. L., Zhu, H. W., et al. (2011). Achieving High Efficiency Silicon-Carbon Nanotube Heterojunction Solar Cells by Acid Doping. *Nano Lett.*, 11(5), 1901-1905.

[6] Hatzikraniotis, E. (2012). On the Recovery of Wasted Heat Using a Commercial Thermoelectric Device. *Acta Phys Pol A.*, 121(1), 287-289.

[7] Kunadian, I., Andrews, R., Menguc, M. P., & Qian, D. (2009). Thermoelectric power generation using doped MWCNTs. *Carbon.*, 47(3), 589-601.

[8] Van Thanh, D., Dzung, Viet. D., Takeo, Y., Bui, Thanh. T., Kenji, H., & Susumu, S. (2010). Integration of SWNT film into MEMS for a micro-thermoelectric device. Smart Materials and Structures. , 19(7), 075003.

[9] Oh, S. H., Kim, K., & Kim, H. (2011). Polypyrrole-modified hydrophobic carbon nanotubes as promising electrocatalyst supports in polymer electrolyte membrane fuel cells. *International Journal of Hydrogen Energy*, 36(18), 11564-11571.

[10] Wang, C., Waje, M., Wang, X., Tang, J. M., Haddon, R. C., & Yan, Y. S. (2004). Proton exchange membrane fuel cells with carbon nanotube based electrodes. *Nano Lett.*, 4(2), 345-348.

[11] Girishkumar, G., Vinodgopal, K., & Kamat, P. V. (2004). Carbon nanostructures in portable fuel cells: Single-walled carbon nanotube electrodes for methanol oxidation and oxygen reduction. *J Phys Chem B.*, 108(52), 19960-19966.

[12] Li, L., Wu, G., & Xu, B. Q. (2006). Electro-catalytic oxidation of CO on Pt catalyst supported on carbon nanotubes pretreated with oxidative acids. *Carbon*, 44(14), 2973-2983.

[13] Acquah, S. F. A., Ventura, D. N., & Kroto, H. W. (2011). Strategies To Successfully Cross-link Carbon Nanotubes. *Electronic Properties of Carbon Nanotubes: InTech.*

[14] Ventura, D. N., Stone, R. A., Chen, K. S., Hariri, H. H., Riddle, K. A., Fellers, T. J., Yun, C. S., Strouse, G. F., Kroto, H. W., et al. (2010). Assembly of cross-linked multi-walled carbon nanotube mats. *Carbon.*, 48(4), 987-994.

[15] Yao, Y. (2012). Hydrogen Storage Using Carbon Nanotubes. In: Marulanda JM, ed., *Carbon Nanotubes*, 543-562.

[16] Li, N., Ma, Y. F., Wang, B., Huang, Y., Wu, Y. P., Yang, X., & Chen, Y. S. (2011). Synthesis of semiconducting SWNTs by arc discharge and their enhancement of water splitting performance with TiO2 photocatalyst. *Carbon*, 49(15), 5132-5141.

[17] Ljeri, V., Cappelletto, L., Bianco, S., Tortello, M., Spinelli, P., & Tresso, E. (2010). Nafion and carbon nanotube nanocomposites for mixed proton and electron conduction. *Journal of Membrane Science*, 363(1-2), 265-270.

[18] Kong, J., Chapline, M. G., & Dai, H. J. (2001). Functionalized carbon nanotubes for molecular hydrogen sensors. *Advanced Materials.*, 13(18), 1384-1386.

[19] Huang, B. R., & Lin, T. C. (2011). A novel technique to fabricate horizontally aligned CNT nanostructure film for hydrogen gas sensing. *International Journal of Hydrogen Energy.*, 36(24), 15919-15926.

[20] Lin, T. C., & Huang, B. R. (2012). Palladium nanoparticles modified carbon nanotube/ nickel composite rods (Pd/CNT/Ni) for hydrogen sensing. *Sensors and Actuators B-Chemical.*, 162(1), 108-113.

[21] Li, W., Hoa, N. D., & Kim, D. (2010). High performance carbon nanotube hydrogen sensor. *Sensors and Actuators B-Chemical.*, 149(1), 184-188.

[22] Jo, Y., Cheon, J. Y., Yu, J., Jeong, H. Y., Han-H, C., Jun, Y., & Joo, S. H. (2012). Highly interconnected ordered mesoporous carbon-carbon nanotube nanocomposites: Pt-free, highly efficient, and durable counter electrodes for dye-sensitized solar cells. *Chemical Communications.*

[23] Chen, T., Qiu, L., Cai, Z., Gong, F., Yang, Z., Wang, Z., & Peng, H. (2012). Intertwined Aligned Carbon Nanotube Fiber Based Dye-Sensitized Solar Cells. *Nano Lett.*, 12(5), 2568-2572.

[24] Bi, H., Huang, F. Q., Liang, J., Tang, Y. F., Lu, X. J., Xie, X. M., & Jiang, M. H. (2011). Large-scale preparation of highly conductive three dimensional graphene and its applications in CdTe solar cells. *J Mater Chem.*, 21(43), 17366-17370.

[25] Leubner, S., Katsukis, G., & Guldi, D. M. (2012). Decorating polyelectrolyte wrapped SWNTs with CdTe quantum dots for solar energy conversion. *Faraday Discuss.*, 155, 253-265.

[26] Bourdo, S. E., Saini, V., Piron, J., Al-Brahim, I., Boyer, C., Rioux, J., Bairi, V., Biris, A. S., & Viswanathan, T. (2012). Photovoltaic Device Performance of Single-Walled Carbon Nanotube and Polyaniline Films on n-Si: Device Structure Analysis. *ACS applied materials & interfaces.*, 4(1), 363-368.

[27] Mattoso, L. H. C., Manohar, S. K., Macdiarmid, A. G., & Epstein, A. J. (1995). Studies on the Chemical Syntheses and on the Characteristics of Polyaniline Derivatives. *Journal of Polymer Science Part a-Polymer Chemistry.*, 33(8), 1227-1234.

[28] Kozawa, D., Hiraoka, K., Miyauchi, Y., Mouri, S., & Matsuda, K. (2012). Analysis of the Photovoltaic Properties of Single-Walled Carbon Nanotube/Silicon Heterojunction Solar Cells. *Appl Phys Express*, 5(4).

[29] Castrucci, P., Del Gobbo, S., Camilli, L., Scarselli, M., Casciardi, S., Tombolini, F., Convertino, A., Fortunato, G., & De Crescenzi, M. (2011). Photovoltaic Response of Carbon Nanotube-Silicon Heterojunctions: Effect of Nanotube Film Thickness and Number of Walls. *J Nanosci Nanotechnol.*, 11(10), 9202-9207.

[30] Hewitt, C. A., Kaiser, A. B., Roth, S., Craps, M., Czerw, R., & Carroll, D. L. (2012). Multilayered Carbon Nanotube/Polymer Composite Based Thermoelectric Fabrics. *Nano Lett.*, 12(3), 1307-1310.

Chapter 8

The Role of Carbon Nanotubes in Enhancement of Photocatalysis

Tawfik A. Saleh

Additional information is available at the end of the chapter

1. Introduction

The chemical, physical and mechanical properties of carbon nanotubes (CNTs) have stimulated extensive investigation since their discovery in the early 1990s (Iijima, 1991). CNTs, which are considered quasi-one dimensional nanostructures, are graphite sheets rolled up into cylinders with diameters of the order of a few nanometers and up to some millimeters in length. Types of nanotubes are the single-walled nanotubes (SWCNTs), double-walled nanotubes (DWCNTs) and the multi-walled nanotubes (MWCNTs). The MWCNTs consist of multiple layers of graphite arranged in concentric cylinders.

During the early stage, the primary research interests include the synthesis or growth of CNTs to prepare enough amounts of CNTs with desired dimension and purity. Several methods like arc discharge, laser ablation of graphite, the more productive chemical vapor deposition (CVD) and plasma enhanced CVD method, have been used to prepare high purity CNTs with controllable wall-thickness and length and acceptable price (Meyyappan, 2004). CNTs attract considerable attention due to their special structure and high mechanical strength which makes them to be good candidates for advanced composites. They can be either semiconducting, semimetallic or metallic, depending on the helicity and the diameter of the tube (Ebbesen et al., 1996; Yang et al., 2003). Based on the structure and shape, CNTs conduct electricity due to delocalization of the pi bond electrons. On the other side, researchers found that CNTs are efficient adsorbents due to their large specific surface area, hollow and layered structures and the presence of pi bond electrons on the surface. Besides that, more active sites can be created on the nanotubes. Thus, CNTs can be used as a promising material in environmental cleaning.

Photocatalytic oxidation using a semiconductor such as TiO_2, ZnO and WO_3 as photocatalyst is one of the advanced oxidation processes used for degradation of various pollutants in industrial wastewaters. As the semiconductor is illuminated with photons having energy content equal to or higher than the band gap, the photons excite valence band (VB) electrons across the band gap into the conduction band(CB), leaving holes behind in the valence band.Thus, there must be at least two reactions occurring simultaneously: oxidation from photogenerated holes, and reduction from photogenerated electrons.

The holes react with water molecules or hydroxide ions (OH$^-$) producing hydroxyl radicals ($^\bullet$OH). The generation of such radicals depends on the pH of the media. Targeted pollutants which are adsorbed on the surface of the catalyst will then be oxidized by $^\bullet$OH. On the other hand, the excited electrons (e$^-$) to the conduction band (CB) can generate hydroxyl radical ($^\bullet$OH) and can also react with O_2 and trigger the formation of very reactive superoxide radical ion ($O_2^{-\bullet}$) that can oxidize the target.

The band gap is characteristic for the electronic structure of a semiconductor and is defined as the energy interval (ΔE_g) between the VB and CB (Koci et al.,2011). VB is defined as the highest energy band in which all energy levels are occupied by electrons, whereas CB is the lowest energy band without electrons. The rate of a photo catalytic reaction depends on several parameters. First and most important is the type of the photo catalytic semiconductor. The second factor is the light radiation used or the stream of photons, as over supply of light accelerates electron–hole recombination (Koci et al.,2008). Third factor is pH of the medium with which the semiconductor surface is in contact with the targeted molecules. Fourth factor is the concentration of the substrate influencing the reaction kinetics. Fifth parameter is the temperature of the media where higher temperatures cause frequent collision between the semiconductor and the substrate (Koci et al.,2010).

The degradation rate can be enhanced by reducing the electron-hole recombination rate; preventing the particles agglomeration; and increasing the adsorption capacity, as it is a key process in the photocatalysis. In order to improve the photocatalytic efficiency, several methods have been investigated. This includes:

1. increasing the surface area of the metal oxide by synthesizing nano-size materials;

2. generation of defect structures to induce space-charge separation and thus reduce the recombination;

3. modification of the semiconductors with metal or other semiconductor; and

4. adding a co-sorbent such as silica, alumina, zeolite or clay (Yu et al. 2002; Rusu and Yates, 1997)

CNTs based composites have attracted considerable attentions due to the intrinsic properties that have been created owing to the addition of CNTs into the composite. Functionalization of CNTs, or attachment of individual atoms, molecules or their aggregates to CNTs, further extend the field of application of these nanosystems in different fields like in photocatalysis process (Dresselhaus & Dresselhaus, 2001; Burghard, 2005; Saleh, 2011). CNT/Metal oxide composites have been recently reported to be used for the treatment of contaminated water. In this chapter, therefore, the application of CNTs to enhance the photocatalytic activity of TiO_2, ZnO and WO_3 will be discussed.

2. Synthesis of carbon nanotube/catalyst composites

There are two main steps for the synthesis of CNT/catalyst nanocomposites. The first step is the grafting of oxygen-containing groups on the surface of the nanotubes and the second step is the attachment of the metal oxides on the active surface of the nanotubes.

2.1. Grafting of oxygen-containing groups on CNTs

Grafting of oxygen-containing groups on the surface of the nanotubes or activation of CNTs can be achieved by oxidation treatment. It can be performed using oxidizing agents such as nitric acid, sulfuric acid, or a mixture of both. For example, oxygen-containing groups can be grafted on the surface of the nanotubes by the following procedure. Initially, CNTs are dispersed by sonication in concentrated acidic media. Then, the mixture is treated by reflux while stirring vigorously at temperature of 100-120°C. After refluxing process, the mixture is allowed to cool at room temperature. The oxidized CNTs are purified by extraction from the residual acids by repeated cycles of dilution with distilled water, centrifugation and decanting the solutions until the pH is approximately 5-6. After the purification process, the oxidized CNTs are dried overnight in an oven at 100°C. After that, the dry oxidized CNTs are pulverized in a ball-mill.

The presence of oxygen containing groups on the surface of the oxidized nanotubes are characterized by the means of Fourier transform infrared spectroscopy (FT-IR), X-ray powder diffraction (XRD), field emission scanning electron microscopy (FESEM) and the transmission electron microscopy (TEM).

As an example, IR spectra, in the range of 400-4000 cm^{-1}, were recorded in KBr pellets using a Thermo Nicolet FT-IR spectrophotometer at room temperature. Samples were prepared by gently mixing 10 mg of each sample with 300 mg of KBr powder and compressed into discs at a force of 17 kN for 5 min using a manual tablet presser. Figure 1 depicts IR spectrum of oxidized MWCNTs. In the spectrum, a characteristic peak at 1580 cm^{-1} can be assigned to C=C bond in MWCNTs. The band at about 1160cm^{-1} is assigned to C–C bonds. Also, the spectrum shows the carbonyl characteristic peak at 1650 cm^{-1}, which is assigned to the carbonyl group from quinine or ring structure. More characteristic peak to the carboxylic group is the peak at 1720 cm^{-1} (Ros et al., 2002; Yang et al., 2005; Xia et al., 2007). The observation of IR spectra corresponding to the oxidized MWCNTs suggests the presence of carboxylic and hydroxylic groups on the nanotube surface.

Figure 2 depicts the typical XRD pattern of the oxidized MWCNTs. The strongest diffraction peak at the angle (2θ) of 25.5° can be indexed as the C(002) reflection of the hexagonal graphite structure (Rosca et al., 2005; Saleh et al., 2011; Lu et al., 2008). The sharpness of the peak at the angle (2θ) of 25.5° indicates that the graphite structure of the MWCNTs was acid-oxidized without significant damage since any decrease in the order of crystallinity in

MWCNTs will make the XRD peaks broader and shift the peak diffraction towards lower angles. The other characteristic diffraction peaks of graphite at 2θ of about 43°, 53° and 77° are associated with C(100), C(004) and C(110) diffractions of graphite, respectively.

Energy dispersive X-ray spectroscope (EDX) measurement is also used as a quantitative analysis for the presence of the oxygen containing groups on the surface of the nanotubes. Figure 4 represents the results of the oxidized MWCNTs. The results shows the presence of oxygen in the sample in addition to carbon element. SEM and TEM are used to characterize the morphology of the nanotube and to ensure that the structure of the nanotube has not been destroyed by the acid treatment. As an example, SEM image and the inset TEM image in Figure 3 confirm that there is no damage effect on the nanotubes using mixtures of nitric acid sulfuric acid for the treatment of the nanotubes.

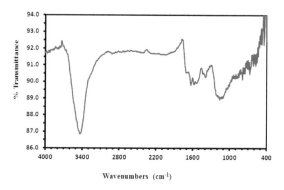

Figure 1. FTIR spectrum of MWCNT oxidized with H_2SO_4/HNO_3 mixture for 6h at 100°C.

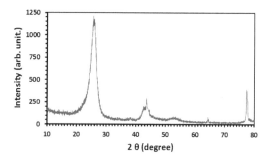

Figure 2. XRD patterns of MWCNT oxidized with H_2SO_4/HNO_3 mixture for 6h at 100°C.

Figure 3. Field emission scanning electron microscopy (FESEM) image of the MWCNTs oxidized with H₂SO₄/HNO₃ mixture for 6h at 100°C; Inset is the transmission electron microscopy (TEM) image of the same.

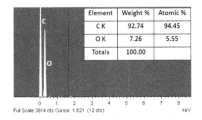

Element	Weight %	Atomic %
C K	92.74	94.45
O K	7.26	5.55
Totals	100.00	

Full Scale 3814 cts Cursor: 1.621 (12 cts)

Figure 4. EDX spectrum of the MWCNTs oxidized with H₂SO₄/HNO₃ mixture for 6h at 100°C; inset is the table showing the percentage of each component in the nanotubes.

2.2. Synthesis of CNT/catalysts nanocomposites

CNT/metal oxide nanocomposites can be synthesized by different methods which fall into two basic classes. The first class involves the prior synthesis of nanoparticles that subsequently connected to surface functionalized CNTs by either covalent or noncovalent interactions (Eder, 2010; Peng et al., 2010; Hu et al., 2010). The second class is the one step method which involves direct deposition of nanoparticles onto MWCNT surface, in situ formation of nanoparticles through redox reactions or electrochemical deposition on CNTs (Chen et al., 2006; Gavalas et al., 2001; Yang et al., 2010; Sahoo et al., 2001; Lee et al., 2008). The second class has the advantages where uniform nanomaterials can be prepared due to the presence of active sites on oxidized CNT surfaces.

As an example, CNT/ZnO nanocomposites are prepared by the following procedure (Saleh et al., 2010). Zinc precursor like Zn(NO₃)₂.6H₂O, is dissolved in doubly deionized water. Then, ammonia is added drop-wise under continuous stirring into the solution to form a clear solution. The oxidized MWCNTs is added into the solution. The mixture is refluxed at 100°C. The composite are separated and dried at 80°C prior to the calcination in vacuum at 300°C.

Different techniques can be applied for the characterization of the nanocomposite. For example XRD is employed to determine crystalline phases and average crystalline size. FT-IR is used for qualitative analysis of the binding of the metal oxide into the nanotube surface. The morphology of the nanotubes and particle size are examined by the field emission scanning electron microscope (FESEM) and high resolution transmission electron microscopy (HRTEM). EDX measurement is also used as a quantitative analysis for the presence of the oxygen containing groups on the surface of the nanotubes. As an example, Figure 5 depicts the EDX data of the CNT/ZnO nanocomposite. The table shows the percentage of each component in the composite. Figure 6, SEM image and the inset HRSEM image, confirm the presence of zinc oxide particles on the surface of the nanotubes.

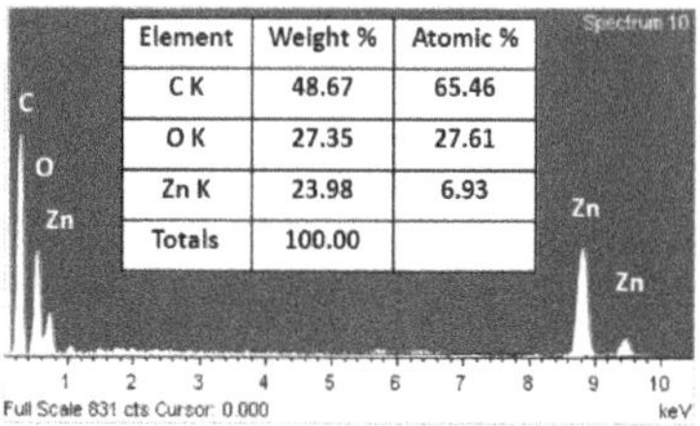

Element	Weight %	Atomic %
C K	48.67	65.46
O K	27.35	27.61
Zn K	23.98	6.93
Totals	100.00	

Figure 5. EDX spectrum of the MWCNT/ZnOnanocomposites; inset is the table showing the percentage of each component in the nanotubes.

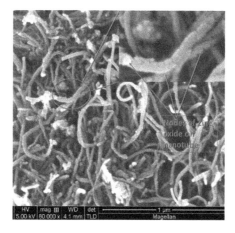

Figure 6. Field emission scanning electron microscopy (FESEM) image of the MWCNT/ZnO; Inset is the HRSEM image.

3. Applications of CNT/Catalyst nanocomposites

CNTs are considered to be good support material for catalysts, because they provide large surface area support and also stabilize charge separation by trapping electrons transferred from metal oxides, thereby hindering charge recombination.

A significant number of papers have been published on the application of CNTs in conjunction with TiO_2, reflecting the focus of recent research (Jitianu et al., 2004; Huang and Gao, 2003; Woan et al., 2009; Feng et al., 2005). One of the most important applications of such composite is to act as photocatalyst for some chemical reactions, especially for the decontamination of organic pollutants in waste waters. The photocatalytic activity of $MWCNT/TiO_2$ composite toward the degradation of acetone under irradiation of UV light was investigated by the detection of the hydroxyl radical ($^{\bullet}OH$) signals using electron paramagnetic resonance. It has been reported that the agglomerated morphology and the particle size of TiO_2 in the composites change in the presence of CNTs, which provide a large surface area resulting in more hydroxyl group on the surface of the composite with no effect on the mesoporous nature of the TiO_2. The composite have been reported to be of higher photocatalytic activity than commercial photocatalyst (P25) and TiO_2/activated carbon (AC) composite (Yu et al., 2005a,b).

The photocatalytic activities of $MWCNT/TiO_2$ under visible light have also been reported using the decolorization of dyes like methylene blue, methyl orange, azo dye and other dyes in model aqueous solutions (Cong et al., 2011; Gao et al., 2009; Hu et al., 2007; Saleh and Gupta, 2012; Yu et al., 2005; Kuo, 2009). TiO_2 loading of 12% was found to result in the highest photoactivity in comparison with 6% and 15% loadings. Little TiO_2 or excessive nanotubes addition shielded the TiO_2 and reduced the UV intensity, due to photon scattering by the nanotubes. However, a high TiO_2 content was found to be ineffective in suppressing exciton recombination because of the large distance between the titania and the nanotubes (Li et al., 2012). Optimum ratio of titania and nanotubes provides a large surface area support and stabilize charge separation by trapping electrons transferred from TiO_2, thereby hindering charge recombination with minimum photon scattering. The composite provides high surface area which is beneficial for photocatalytic activity, as it provides high concentration of target organic substances around sites activated by ultraviolet (UV) radiation.

Also, the activity of $MWCNT/TiO_2$ composites has been investigated in photodegradation of phenol and photocatalytic oxidation of methanol under irradiation of visible light (Wang et al., 2005; An et al., 2007; Yao et al., 2008; Dechakiatkrai et al., 2007). The catalysts exhibited enhanced photocatalytic activity for degradation of toluene in gas phase under both visible and simulated solar light irradiation compared with that of commercial Degussa P25 (Wu et al., 2009). It exhibited high activity for the photoreduction of Cr(VI) in water (Xu et al., 2008). Its efficiency was higher compared to a mechanical mixture of TiO_2 and MWCNTs. A probable synergistic effect of TiO_2 and MWCNTs in a composite $MWCNT/TiO_2$ on the enhancement of visible light performance, have been proposed where MWCNTs act as support, absorbent, photo-generated transfer station and carbon-doping source to narrow the band gap of TiO_2.

The composite has been reported for photoinactivation of E. coli in visible light irradiation (Akhavan et al., 2009). The efficiency of the nanocomposite was high toward photocatalytic hydrogen generation and for the reduction of CO_2 with H_2O (Dai et al., 2009; Xia et al., 2007).

Zinc oxide, a direct wide band gap (3.37 eV) semiconductor with a large excitation binding energy (60 meV), has been investigated as a potential non-toxic photocatalyst used to successfully degrade organic pollutants. Recently, ZnO nanoparticles have received much attention due to its high photoactivity in several photochemical, UV light response, photoelectron-chemical processes and its low cost production possibility (Wu et al., 2008; Neudeck et al., 2011; Gondal et al., 2010; Drmosh et al., 2010).

Experimental results proved that CNT/ ZnO nanocomposites display relatively higher photocatalytic activity than ZnO nanoparticles for the degradation of some dyes like rhodamine B, azo-dyes, methylene blue, methylene orange (Dai et al., 2012; Zhu et al., 2009). The complete removal of azo-dyes such as acid orange, acid bright red, acid light yellow, after selection of optimum operation parameters such as the illumination intensity, catalyst amount, initial dye concentration and the different structures of the dye on the photocatalytic process, can be achieved in relatively short time by using CNT/ZnO composites.

The MWCNT/ZnO nanocomposites exhibits excellent photocatalytic activity toward other pollutants such as acetaldehyde and cyanide in model solutions (Saleh et al., 2011; Saleh et al., 2010). CNTs act as a photogenerated electron acceptor and retard the recombination of photoinduced electron and hole. The adsorption and photocatalytic activity tests indicate that the CNTs serve as both an adsorbent and a visible light photocatalyst. The experimental results show that the photocatalytic activity of the ZnO/MWCNTs nanocomposites strongly depends on the synthetic route, which is probably due to the difference of surface states resulted from the different preparation processes (Zhang, 2006; Kim and Sigmund, 2002; Jiang and Gao, 2005;Agnihotri et al., 2006).

CNT/WO_3 nanocomposites have been synthesized via different routs (Pietruszka et al., 2005; Wang et al., 2008; Saleh and Gupta, 2011). The utilization of carbon nanotubes to enhance photocatalytic activity of tungsten trioxide has also been investigated. The photocatalytic activities are greatly improved when CNT/WO_3 nanocomposite has been used for the degradation of pollutants such as rhodamine B under ultraviolet lamp or under sunlight. The results showed that photocatalytic activity of the $MWCNT/WO_3$ composites prepared by chemical process is higher than that prepared by mechanical mixing. The photocatalytic activity is enhanced when WO_3 nanoparticles are loaded on the surface of CNTs. The enhanced photocatalytic activity may be ascribed to the effective electron transfer between nanotubes and the metal nanoparticles.

A possible synergistic effect between the semiconductor nanoparticles and CNTs on the enhancement of photocatalytic activity is proposed in Figure 7. The mechanism is based on the results of the structure characterizations and the enhancement in photocatalytic activity of the prepared composite.

When the catalyst is irradiated by photons, electrons (e^-) are excited from the valence band (VB) to the conduction band (CB) of catalysts or the metal oxide nanoparticles (NP) creating a charge vacancy or holes (h⁺), in the VB. Some of the charges quickly recombine without creating efficient photodecomposition of the pollutant. In the case where the composite is applied, the strong interaction between the nanotube and the metal oxide results in a close contact to form a barrier junction which offers an effective route of reducing electron-hole recombination by improving the injection of electrons into the nanotube. Therefore, CNTs acts as a photo-generated electron acceptor to promote interfacial electron transfer process since CNTs are relatively good electron acceptor while the semiconductor is an electron donor under irradiation (Saleh and Gupta, 2011; Riggs et al., 2000; Subramanian et al., 2004; Geng et al., 2008). The adsorbed oxygen molecules on the nanotubes react with the electrons forming very reactive superoxide radical ion (O2⁻) which oxidize the target. On the other side, the hole (h⁺) oxidize hydroxyl groups to form hydroxyl radical (˙OH) which can decompose the target.

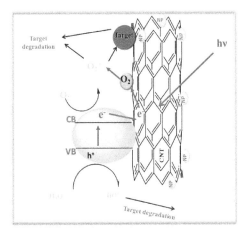

Figure 7. Schematic diagram of the proposed mechanism of photodegradation over CNT/MO composite.

Some important points of such process can be highlighted as:

- Stronger adsorption on photocatalyst for the targeted moleculs of pollutant is achieved by the incorporation of the nanotubes, due to their large specific surface area and high quality active sites.

- The nanotubes can act as effective electron transfer unitbecause of their high electrical conductivity and high electron storage capacity.

- The nanotubes manifest higher capture electron ability and can prompt electron transfer from the conduction band of the metal oxide or semiconductor nanoparticles (NP) towards the nanotube surface due to their lower Fermi level (Cong et al., 2011).

- Schottky barrier forms at the interface between the CNTs and the semiconductor. The photo-generated electrons may move freely towards the CNT surface, thus the left holes may move to the valence band (Woan et al., 2009; Chen et al., 2005).

- The presence of the nanotubes in the composite can inhibit the recombination of photo-generated electrons and holes, thus, improving the photocatalytic activity.

- The transmission stability of promoted electron between the nanotubes and the conduction band is enhanced by the strong interaction and intimate contact between the nanoparticles and the surface of the nanotubes.

4. Conclusion

The chapter discusses the preparation of the nanocomposites consisting of carbon nanotubes and metal oxides like titania, zinc oxide and tungsten trioxide. For the preparation of such composite, the oxygen-containing groups are grafted on the surface of the nanotubes by acid treatment. This is followed by the attachment of the metal oxides nanoparticles on the nanotubes surface. The chapter also highlights the means by which the composite is characterized. These include Fourier transform infrared spectroscope, X-ray powder diffraction, field emission scanning electron microscope, energy dispersive X-ray spectroscope and transmission electron microscope.

The UV, visible light and sunlight photocatalytic activity of the CNT-based nanocomposites is higher than that of the metal oxide or mechanical mixture of the metal oxide and CNTs. CNTs are considered to be good support materials for semiconductors like TiO_2, ZnO and WO_3 because nanotubes provide a large surface area support with high quality active sites. Also they stabilize charge separation by trapping electrons, thereby hindering electron–hole recombination by modification of band-gap and sensitization.

Acknowledgements

The author would like to acknowledge the support of Chemistry Departments, Center of Research Excellence in Nanotechnology & King Fahd University of Petroleum and Minerals, (KFUPM) Dhahran, Saudi Arabia, for this work.

Author details

Tawfik A. Saleh*

Address all correspondence to: tawfik@kfupm.edu.sa

Chemistry Department, Center of Excellence in Nanotechnology, King Fahd University of Petroleum & Minerals, Saudi Arabia

References

[1] Agnihotri, S., Mota, J. P. B., Rostam-Abadi, M., & Rood, M. J. (2006). Adsorption site analysis of impurity embedded single-walled carbon nanotube bundles. *Carbon*, 44(1), 2376-2383.

[2] Akhavan, O., Abdolahad, M., Abdi, Y., & Mohajerzadeh, S. (2009). Synthesis of titania/carbon nanotube heterojunction arrays for photoinactivation of E. coli in visible light irradiation. *Carbon*, 47(14), 3280-3287.

[3] An, G., Ma, W., Sun, Z., Liu, Z., Han, B., Miao, S., Miao, Z., & Ding, K. (2007). Preparation of titania/carbon nanotube composites using supercritical ethanol and their photocatalytic activity for phenol degradation under visible light irradiation. *Carbon*, 45, 1795-1801.

[4] Burghard, M. (2005). Electronic and vibrational properties of chemically modified singlewall carbon nanotubes. *Surf. Sci. Rep.*, 0167-5729, 58(1-4), 1-109.

[5] Chen, W., Pan, X., Willinger, M. G., Su, D. S., & Bao, X. (2006). Facile autoreduction of iron oxide/carbon nanotube encapsulates. *J. Am. Chem. Soc.*, 128, 3136-3137.

[6] Chen, Y., Crittenden, J. C., Hackney, S., Sutter, L., & Hand, H. W. (2005). Preparation of a Novel TiO2-Based p–n Junction Nanotube Photocatalyst. *Environ. Sci. Technol.*, 39(5), 1201-1208.

[7] Cong, Y., Li, X., Qin, Y., Dong, Z., Yuan, G., Cui, Z., & Lai, X. (2011). Carbon-doped TiO_2 coating on multiwalled carbon nanotubes with higher visible light photocatalytic activity. *Applied Catalysis B: Environmental*, 107, 128-134.

[8] Dai, K., Dawson, G., Yang, S., Chen, Z., & Lu, L. (2012). Large scale preparing carbon nanotube/zinc oxide hybrid and its application for highly reusable photocatalyst. *Chemical Engineering Journal*, 191, 571-578.

[9] Dai, K., Peng, T., Ke, D., & Wei, B. (2009). Photocatalytic hydrogen generation using a nanocomposite of multi-walled carbon nanotubes and TiO_2 nanoparticles under visible light irradiation. *Nanotechnology*, 20(12), 125603.

[10] Dechakiatkrai, C., Chen, J., Lynam, C., Phanichphant, S., & Wallace, G. G. (2007). Photocatalytic oxidation of methanol using titanium dioxide/single-walled carbon nanotube composite. *J Electrochem Soc.*, 154(5), A407-411.

[11] Dresselhaus, M. S., & Dresselhaus, G. (2001). Carbon Nanotubes: Synthesis, Structure, Properties and Applications: Topics in Applied Physics, Springer-Verlag. ISBN 3-54041-086-4, Berlin.

[12] Drmosh, Q. A., Gondal, M. A., Yamani, Z. H., & Saleh, T. A. (2010). Spectroscopic Characterization Approach to Study Surfactants Effect On ZnO Nanoparticles Synthesis by Laser Ablation Process. *Applied Surface Science*, 256, 4661-4666.

[13] Ebbesen, T. W., Lezec, H. J., Hiura, H., Bennett, J. W., Ghaemi, H. F., & Thio, T. (2010). Electrical conductivity of individual carbon nanotubes. *Nature*, 382, 6586 (1996) 54-56.

[14] Eder, D. Carbon nanotube-inorganic hybrids. *Chem. Rev.*, 110, 1348-1385.

[15] Feng, W., Feng, Y., Wu, Z., Fujii, A., Ozaki, M., & Yoshino, K. (2005). Optical and electrical characterizations of nanocomposite film of titania adsorbed onto oxidized multiwalled carbon nanotubes. *J Phys Condens Matter*, 17(27), 4361-4368.

[16] Gao, B., Chen, G. Z., & Li, P. G. (2009). Carbon nanotubes/titanium dioxide (CNTs/TiO$_2$) nanocomposites prepared by conventional and novel surfactant wrapping sol-gel methods exhibiting enhanced photocatalytic activity. *Appl Catal*, B89(3-4), 503-509.

[17] Gavalas, V. G., Andrews, R., Bhattacharyya, D., & Bachas, L. G. (2001). Carbon nanotube sol-gel composite materials. *Nano Lett.*, 1, 719-721.

[18] Geng, Q., Guo, Q., Cao, C., & Wang, L. (2008). Investigation into NanoTiO$_2$/ACSPCR for Decomposition of Aqueous Hydroquinone. *Ind. Eng. Chem. Res.*, 47, 2561-2568.

[19] Gondal, M. A., Drmosh, Q. A., Yamani, Z. H., & Saleh, T. A. (2010). Effect of post-annealing temperature on structural and optical properties of nano-ZnO synthesized from ZnO$_2$ by Laser Ablation Method. *International Journal of NanoParticles*, 3(3), 257-266.

[20] Hu, G., Meng, X., Feng, X., Ding, Y., Zhang, S., & Yang, M. (2007). Anatase TiO$_2$ nanoparticles/carbon nanotubes nanofibers: preparation, characterization and photocatalytic properties. *J Mater Sci*, 42(17), 7162-7170.

[21] Hu, L., Hecht, D. S., & Gruner, G. (2010). Carbon nanotube thin films: fabrication, properties, and applications. *Chem. Rev.*, 110, 5790-5844.

[22] Huang, Q., & Gao, L. (2003). Immobilization of rutile TiO$_2$ on multiwalled carbon nanotubes. *J Mater Chem*, 13(7), 1517-9.

[23] Iijima, S. (1991). Helical Microtubules of Graphitic Carbon. *Nature*, 354, 56-58.

[24] Jiang, L., & Gao, L. (2005). Fabrication and characterization of ZnO-coated multi-walled carbon nanotubes with enhanced photocatalytic activity. *Mater. Chem. Phys.*, 91, 313-316.

[25] Jitianu, A., Cacciaguerra, T., Benoit, R., Delpeux, S., Beguin, F., & Bonnamy, S. (2004). Synthesis and characterization of carbon nanotubes-TiO$_2$ nanocomposites. *Carbon*, 42(5-6), 1147-1151.

[26] Kim, H., & Sigmund, W. (2002). Zinc oxide nanowires on carbon nanotubes. *Appl. Phys. Lett.*, 81, 2085-2088.

[27] Koci, K., Mateju, K., Obalova, L., Krejcikova, S., Lacny, Z., Placha, D., Capek, L., Hospodkova, A., & Solcova, O. (2010). Effect of silver doping on the TiO$_2$ for Photocatalytic reduction of CO$_2$. *Applied Catalysis B: Environmental*, 96, 239-244.

[28] Koci, K., Obalová, L., & Lacný, Z. (2008). Photocatalytic reduction of CO$_2$ over TiO$_2$ based catalysts. *Chemical Papers*, 62, 1-9.

[29] Koci, K., Reli, M., Kozák, O., Lacny, Z., Plachá, D., Praus, P., & Obalov, I. (2011) Influence of reactor geometry on the yield of CO2 Photocatalytic reduction. *Catalysis Today*, 176(1), 212-214.

[30] Kuo, C-Y. (2009). Prevenient dye-degradation mechanisms using UV/TiO$_2$/carbon nanotubes process. *J Hazard Mater*, 163(1), 239-244.

[31] Lee, D. H., Park, J. G., Choi, K. J., Choi, H. J., & Kim, D. W. (2008). Preparation of brookite-Type TiO$_2$/carbon nanocomposite electrodes for application to Li ion batteries. *Eur. J. Inorg. Chem.*, 6, 878-882.

[32] Li, Y., Leiyong, Li., Chenwan, Li., Chen, W., & Zeng, M. (2012). Carbon nanotube/ titania composites prepared by a micro-emulsion method exhibiting improved photocatalytic activity. *Applied Catalysis A: General*, 427(428), 1-7.

[33] Lu, C., Su, F., & Hu, S. (2008). Surface modification of carbon nanotubes for enhancing BTEX adsorption from aqueous solutions. *Applied Surface Science*, 254, 7035-7041.

[34] Meyyappan, M. (2004). (Ed.), *Carbon Nanotubes: Science and Applications*, CRC Press, 0-84932-111-5.

[35] Neudeck, C., Kim, Y. Y., Ogasawara, W., Shida, Y., Meldrum, F., & Walsh, D. (2011). General route to functional metal oxide nanosuspensions, enzymatically deshelled nanoparticles, and their application in photocatalytic water splitting. *Small*, 7, 869-873.

[36] Peng, X., Sfeir, M. Y., Zhang, F., Misewich, J. A., & Wong, S. S. (2010). Covalent synthesis and optical characterization of double-walled carbon nanotube-nanocrystal heterostructures. *J. Phys. Chem. C*, 114, 8766-8773.

[37] Pietruszka, B., Gregorio, F. D., Keller, N., & Keller, V. (2005). High-efficiency WO3/ carbon nanotubes for olefin skeletal isomerization. *Catal. Today*, 102-103, 94-100.

[38] Riggs, J. E., Guo, Z., Carroll, D. L., & Sun, Y. P. (2000). Strong Luminescence of Solubilized Carbon Nanotubes. *J. Am. Chem. Soc.*, 122, 5879-5880.

[39] Ros, T. G., van Dillen, A. J., Geus, J. W., & Koningsberger, D. C. (2002). Surface oxidation of carbon nanofibres. *Chem Eur J*, 8, 1151-1162.

[40] Rosca, I. D., Watari, F., Uo, M., & Akasaka, T. (2005). Oxidation of multiwalled carbon nanotubes by nitric acid. *Carbon*, 43, 3124-31.

[41] Rusu, C. N., & Yates Jr, J. T. (1997). Defect sites on TiO2(110). Detection by O2 photodesorption. *Langmuir*, 13(16), 4311-4316.

[42] Sahoo, S., Husale, S., Karna, S., Nayak, S. K., & Ajayan, P. M. (2011). Controlled as-sembly of Ag nanoparticles and carbon nanotube hybrid structures for biosensing. *J. Am. Chem. Soc.*, 133, 4005-4009.

[43] Saleh, T. A. (2011). The influence of treatment temperature on the acidity of MWCNT oxidized by HNO3 or a mixture of HNO3/H2SO4. *Applied Surface Science*, 257, 17 June, 7746-7751.

[44] Saleh, T. A., & Gupta, V. K. (2011). Functionalization of tungsten oxide into MWCNT and its application for sunlight-induced degradation of rhodamine B. *Journal of Col-loid and Interface Science*, 362(2), 337-344.

[45] Saleh, T. A., & Gupta, Vinod. K. (2012). Photo-catalyzed degradation of hazardous dye methyl orange by use of a composite catalyst consisting of multi-walled carbon nanotubes and titanium dioxide. *Journal of Colloid and Interface Science*, 371(1), 101-106.

[46] Saleh, T. A., Gondal, M. A., Drmosh, Q. A., Z Yamani, H. A., & AL-yamani, A. (2011). Enhancement in photocatalytic activity for acetaldehyde removal by embedding ZnO nano particles on multiwall carbon nanotubes. *Chemical Engineering Journal*, 166(1), 407-412.

[47] Saleh, T. A., Gondal, M. A., & Drmosh, Q. A. (2010). Preparation of a MWCNT/ZnO nanocomposite and its photocatalytic activity for the removal of cyanide from water using a laser. *Nanotechnology*, 21(49), 8, doi:10.1088/0957-4484/21/49/495705.

[48] Subramanian, V., Wolf, E., & Kamat, P. V. (2004). Catalysis with TiO2/Gold Nano-composites. Effect of Metal Particle Size on the Fermi Level Equilibration. *J. AM. CHEM. SOC.*, 126, 4943-4950.

[49] Wang, S., Xiaoliang, Shi., Gangqin, Shao., Xinglong, Duan., Hua, Yang., & Tianguo, Wang. (2008). Preparation, characterization and photocatalytic activity of multi-wal-led carbon nanotube-supported tungsten trioxide composites. *Journal of Physics and Chemistry of Solids*, 69, 2396-2400.

[50] Wang, W. D., Serp, P., Kalck, P., & Faria, J. L. (2005). Visible light photodegradation of phenol on MWCNT-TiO$_2$ composite catalysts prepared by a modified sol-gel method. *J Mol Catal A: Chem*, 235(1-2), 194-9.

[51] Woan, K., Pyrgiotakis, Georgios., & Sigmund, Wolfgang. (2009). Photocatalytic Car-bon-Nanotube-TiO2 Composites. *Advanced Materials*, 21(21), 2233-2239.

[52] Wu, X., Jiang, P., Cai, W., Bai, X. D., Gao, P., & Xie, S. S. (2008). Hierarchical ZnO mi-cro-/nanostructure film. *Adv. Eng. Mater.*, 10, 476-481.

[53] Wu, Z., Fan, Dong., Weirong, Zhao., Haiqiang, Wang., Yue, Liu., & Baohong, Guan. (2009). The fabrication and characterization of novel carbon doped TiO$_2$ nanotubes, nanowires and nanorods with high visible light photocatalytic activity. *Nanotechnolo-gy*, 20(23), 235701.

[54] Xia, W., Wang, Y., Bergstraberr, R., Kundu, S., & Muhler, M. (2007). Surface charac-
      terization of oxygen-functionalized multi-walled carbon nanotubes by high-resolu-
      tion X-ray photoelectron spectroscopy and temperature-programmed desorption.
      Applied Surface Science, 254, 247-250.

[55] Xia, X-H., Jia, Z-J., Yu, Y., Liang, Y., Wang, Z., & Ma, L-L. (2007). Preparation of mul-
      ti-walled carbon nanotube supported TiO_2 and its photocatalytic activity in the re-
      duction of CO_2 with H_2O. *Carbon*, 45(4), 717-21.

[56] Xu, Z., Long, Y., Kang, S-Z., & Mu, J. (2008). Application of the composite of TiO_2
      nanoparticles and carbon nanotubes to the photoreduction of Cr(VI) in water. *J Dis-
      persion Sci Technol*, 29(8), 1150-2.

[57] Yang, C., Wohlgenannt, M., Vardeny, Z. V., Blau, W. J., Dalton, A. B., Baughman, R.,
      & Zakhidov, A. A. (2003). Photoinduced charge transfer in poly(p-phenylene vinyl-
      ene) derivatives and carbon nanotube/C60 composites. *Physica B: Condensed Matter*,
      338(1-4), 366-369.

[58] Yang, D.-Q., Rochette, J-F., & Sacher, E. (2005). Functionalization of multiwalled car-
      bon nanotubes by mild aqueous sonication. *J Phys Chem B*, 109, 7788-7794.

[59] Yang, J., Jiang, L. C., Zhang, W. D., & Gunasekaran, S. (2010). Highly sensitive non-
      enzymatic glucose sensor based on a simple two-step electrodeposition of cupric ox-
      ide (CuO) nanoparticles onto multi-walled carbon nanotube arrays. *Talanta*, 82, 25-33.

[60] Yao, Y., Li, G., Ciston, S., Lueptow, R. M., & Gray, K. A. (2008). Photoreactive TiO_2/
      carbon nanotube composites: synthesis and reactivity. *Environ Sci Technol*, 42(13),
      4952-7.

[61] Yu, J. C., Zhang, L., & Yu, J. (2002). Rapid synthesis of mesoporous TiO_2 with high
      photocatalytic activity by ultrasound-induced agglomeration. *New Journal of Chemis-
      try*, 26(4), 416-420.

[62] Yu, Y., Yu, J. C., Chan, C-Y., Che, Y-K., Zhao, J-C., & Ding, L. (2005a). Enhancement
      of adsorption and photocatalytic activity of TiO2 by using carbon nanotubes for the
      treatment of azo dye. *Appl Catal B*, 61(1-2), 1-11.

[63] Yu, Y., Yu, Jimmy C., Yu, Jia-Guo., Kwok, Yuk-Chun., Che, Yan-Ke., Zhao, Jin-Cai.,
      Ding, Lu, Ge, Wei-Kun, & Wong, Po-Keung. (2005b). Enhancement of photocatalytic
      activity of mesoporous TiO2 by using carbon nanotubes. *Applied Catalysis A: General*,
      289, 186-196.

[64] Zhang, W. D. (2006). Growth of ZnO nanowires on modified well-aligned carbon
      nanotube arrays. *Nanotechnology*, 17, 1036-1040.

[65] Zhu, L.-P., Liao, Gui-Hong, Huang, Wen-Ya, Ma, Li-Li, Yang, Yang, Yu, Ying, & Fu,
      Shao-Yun. (2009). Preparation, characterization and photocatalytic properties of
      ZnO-coated multi-walled carbon nanotubes. *Materials Science and Engineering B*, 163,
      194-198.

Chapter 9

Adsorption of Methylene Blue on Multi-Walled Carbon Nanotubes in Sodium Alginate Gel Beads

Fang-Chang Tsai, Ning Ma, Lung-Chang Tsai,
Chi-Min Shu, Tao Jiang, Hung-Chen Chang,
Sheng Wen, Chi Zhang, Tai-Chin Chiang,
Yung-Chuan Chu, Wei-Ting Chen, Shih-Hsin Chen,
Han-Wen Xiao, Yao-Chi Shu and Gang Chang

Additional information is available at the end of the chapter

1. Introduction

Surface water contamination by pollutants is common in highly industrialized countries due to direct discharge of industrial effluents into bodies of water or precipitation of air-borne pollutants into surface water [Murakamia et. al., 2008]. Dyes from the pollutants released along with industrial effluents are easily detected because of their inherently high visibility, meaning that concentrations as low as 0.005 mg/L can easily be detected and capture the attention of the public and the authorities [Ray et. al., 2003, Ray et. al., 2002]. Apart from the aesthetic problems caused by dyes, the greatest environmental concern with dyes is their absorption and reflection of sunlight entering the water, which interferes with the growth of bacteria, such that bacteria levels are insufficient to biologically degrade impurities in the water [Pierce, 1994. Ledakowicz et. al., 2001]. Methylene blue (MB) and methyl violet are two common dyes that have been shown to induce harmful effects on living organisms during short periods of exposure [Hameed et. al., 2009]. Oral ingestion of MB results in a burning sensation and may cause nausea, vomiting, diarrhea, and gastritis. The accidental consumption of large dose induces abdominal and chest pain, severe headache, profuse sweating, mental confusion, painful micturation, and methemoglobinemia [Yasemin et. al., 2006]. Inhalation of methyl violet may cause irritation to the respiratory tract, vomiting, diarrhea, pain, headaches, and dizziness; long-term exposure may cause damage to the mucous membranes and gastrointestinal tract [Allen & Koumanova, 2005]. The majority of dyes in this class are synthetic and usually composed of aromatic rings, which makes them carcinogenic and mutagenic [Ghosh & Bhattacharyya, 2002. Chen et. al., 2003]; they are inert and non-biodegradable when discharged into waste streams [Mittal & Gupta, 1996, Seshadri et. al., 1994]. With the social and economic development, the environmental consciousness of citizens and governing agencies was enhanced. Environmental pollution issues have garnered a considerable amount of attention throughout the world [Renmin et. al., 2005]. MB is a good representative of organic dyes that are difficult to degrade and substantially damage the environment due to their toxicity and dark color [Ho et. al., 2005].

Since carbon nanotube (CNT) was first discovered by S. Ijima in 1991, it has become an academic research subject of great interest [Olson, 1994]. CNT is the thinnest tubular structure humans can presently fabricate. It is lightweight and has high strength, high toughness, flexibility, high surface area, high thermal conductivity, and good electric conductivity and is chemically stable [Baughman et. al., 2002, Thostenson et. al., 2001, Banerjee et. al., 2005]. To fully exploit the superior mechanical, electrical and optical properties of multi-walled carbon nanotube (MWCNT), dispersion and adhesion to a polymeric matrix is a key issue [Iijima, 1991]. Both the dispersibility and matrix adhesion of MWCNT can be improved either by covalent or noncovalent functionalization. For covalent functionalization, several approaches studied, each having its advantages and drawbacks; examples of such methods include wet chemical methods with typical treatment times of up to 24 h [Sahoo et. al., 2010, Liu et. al., 1998, Chen et. al., 1998], treatment in air at elevated temperatures [Tsai et. al., 2010, Ajayan et. al., 1993], by ozone oxidation [Ago et. al., 1999, Mawhinney et. al., 2000] and treatment with low-pressure plasmas [Simmons et. al., 2006, Tseng et. al., 2006, Chen et. al., 2010, Zschoerper et. al., 2009].

Alginate is a collective term for a family of exopolysaccharides produced mainly by brown seaweeds. It has been widely used in the food, biomedical, pharmaceutical, and sewage-treatment industries, preferentially as sodium alginate due to its solubility in cold water. In molecular terms, alginate is composed of (1–4)-linked b-D-mannuronic acid (M) unbranched binary copolymer and a-L-guluronic acid (G) monomer residue, constituting M-, G-, and MG sequential block structures [Chen et. al., 2009]. Most applications that use alginate are based on its gel-forming ability through cation binding: the transition from water-soluble sodium alginate (SA) to water insoluble calcium alginate, for example. Divalent cations preferentially bind toward the G-block rather than the M-block [Moe et. al., 1995, Braccini et. al., 1999]. The composition of monomers and their sequential character (i.e., blackness) affects the gelatin behavior of alginate. In the presence of Ca^{2+}, G-rich samples generally form hard and brittle gels while M-rich samples from soft and elastic gels [Braccini & Perez, 2001, Courtois et. al., 1993, Thakur et. al., 1997, Pe'rez et. al., 1996]. The "egg-box" model has been accepted as a general model to describe gel formation [Morris et. al., 1978, Thom et. al., 1985]. Alginate is an excellent polymer flocculant and has been widely used in wastewater treatment.

This study reports for the first time the effect of the carboxylation method on CNT structure and property. The results can be used as reference for selecting the carboxylation method. Furthermore, the applicability of a new adsorbent, SA and MWCNT and the SA/MWCNT composite, for the sorption of MB dyes from an aqueous solution were investigated.

2. Experimental

2.1. Materials and methods

The SA, MB, and MWCNT were used as received from Fuchen Chemical Reagents Factory, Tianjin, China and Nanotechnologies Port Co., Ltd., Shenzhen, China. The MWCNTs was treated with a mixture of sulfuric and nitric acid under ultrasonic vibration, as seen in Table 1. According to the series reaction time in Table 1, the optimized ratio of the MWCNT to acid mixture is 3:1 by volume. Ultrasonic treatment was applied for the duration of varying reaction times. Filtration was conducted with a micropore filter and sand core filter. Pure de–ionized water (pure DI water) was used to rinse the filtrate until the pH of the aqueous solution was neutral. The compositions of the SA, MWCNT, and SA/MWCNT series specimens prepared in this study are summarized in Table 2. Ten milliliters of an aqueous solution of SA/MWCNT was added drop-wise to 50 mL of calcium chloride (10%, w/v) aqueous solution for 20 min, followed by the sampling of supernatant at the specified time intervals. The gel particles were pre-consolidated under a pressure of 8–30 kPa in a consolidation cell with an inner diameter of 2.0–3.0 mm to produce a packed gel bed to determine their expression characteristics. The schematic evolution of the SA and MWCNT in the microsphere, as a function of the calcium chloride, is shown in Figure 1. Other supplementary agents were of analytical grade (purity > 99.8 mass%) and all solutions were prepared with double distilled water.

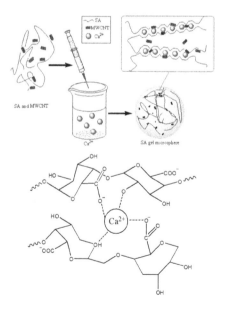

Figure 1. The preparation process of the SA/MWCNT composite gel beads.

A–series reaction group	Mixed-acid treatment time (h)	Hydrogen peroxide treatment time (h)	B–series reaction group	Mixed-acid treatment time (h)	Hydrogen peroxide treatment time (h)
A0	0	0	B0	0	0
A1	1	0	B1	1	0.5
A2	2	0	B2	2	1
A3	4	0	B3	4	2
A4	6	0	B4	6	3
A5	8	0	B5	8	4

Table 1. Formulation for CNT carboxylation.

Sample		SA (%, w/v)	MWCNT (%, w/v)	CaCl$_2$ (%, w/w)
0#	SA$_0$MWCNT$_0$	0	0	0
1#	SA$_2$MWCNT$_0$	2	0	10
2#	SA$_2$MWCNT$_{0.03}$	2	0.03	10
3#	SA$_2$MWCNT$_{0.06}$	2	0.06	10
4#	SA$_2$MWCNT$_{0.09}$	2	0.09	10
5#	SA$_2$MWCNT$_{0.12}$	2	0.12	10
6#	SA$_2$MWCNT$_{0.15}$	2	0.15	10

Table 2. The composites of SA/MWCNT series samples.

2.2. MWCNT dispersed polarity

Six small reagent bottles were filled with 6 mL pure DI water, 4 mL toluene and a small amount of MWCNT derived as shown in Table 1. They were ultrasonically treated for 0.5 h, and then, after the solution was stored for 12 h, they were recovered and observed.

2.3. Particle size analysis

The particle size analysis measurements of MWCNT and modified MWCNT series specimens were recorded using a Dandong Bettersize Instruments Ltd. BT-9300H at 25°C and 50% relative humidity, wherein six scans with a size range of 0.1–340 µm were collected during each data measurement. Particle size analysis samples of powder specimens were collected using approximately 15 mL pure DI water and a small amount of MWCNT derived as shown in Table 1.

2.4. Adsorption property

All sorption measurements were performed by batch type with 50 mL of MB solution in a shaking incubator to form a final concentration of 50 mg/L ($A_{665\,nm}$ = 2.9966) at room temperature for 3 h. The equilibrium MB concentration was measured by means of double beam ultraviolet–visible spectroscopy (Shanghai Precision & Scientific Instrument Co., Ltd, UV762, China), and the pH values of the solution were measured using a pH meter (Shanghai Yulong Instrument Co., Ltd., PHS-3 C, China) with a calomel and glass electrode (E201-9). The dye decolorization percentage was defined as follows:

$$Decoloration\ percentage(\%) = (A_0 - A) / A_0 \times 100\% \tag{1}$$

where A_0 is the dye absorbance of the control specimen, A is the dye absorbance of the reacted sample.

2.5. Electrical conductivity

To understand the electrical conductivity properties of MWCNT in SA specimens dispersed in MB solution, the electrical conductivity of the SA and SA/MWCNT solutions were measured at 25°C and 50% relative humidity using a conductivity meter (LIDA Instrument Factory, DDS-12A, China).

3. Results and discussion

3.1. Carbon nanotube dispersed polarity

A typical photograph of the polarity of MWCNT and modified MWCNT specimens is shown in Fig. 2. Fig. 2 shows the dispersion of the modified MWCNT in aqueous and organic solvent solutions after being exposed to the treatments highlighted in Table 1 and then left undisturbed for 12 h. The figure shows that in the six groups of MWCNT, except the unmodified carbon nanotube, there always exists an interface of two phases that cannot be dissolved in one another. All five of the other groups show different extents of dispersion. MWCNT[b3] shows the most stable dispersion in aqueous phase; even after being aged for a week, it still maintained the state seen in Fig. 2.

Figure 2. Photograph depicting the polarity of pure MWCNT specimens.

3.2. Particle Size Analysis

Particle size analysis was conducted on A–series and B–series of carboxylated MWCNT. Figs. 3 and 4 show that as carboxylation reaction time increases the extent of carbon nanotube shortening is increased; this is particularly true for the B–series, where the mixed-acid, hydrogen peroxide, and ultrasonic treatment times were all shortened. The B–series samples are much shorter than the A–series samples of MWCNT treated only with mixed-acid and ultrasonic treatment. This finding further corroborates the FT-IR results. With longer carboxylation reaction times the MWCNT is more severely damaged, inducing greater rupture on the C–C bond of the CNT. The higher activity at MWCNT ends facilitates bonding with free O and H from water or solution and the formation of carboxyl groups on the fracture site, increasing the functionalized carboxyl groups and the extent of MWCNT carboxylation.

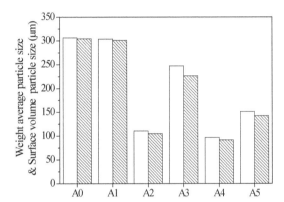

Figure 3. The A–series of MWCNT specimens on weight average particle size (white column) and surface volume particle size (slash column) measured at 25°C.

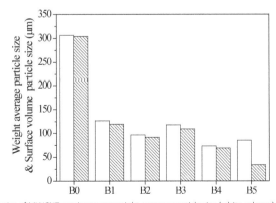

Figure 4. The B–series of MWCNT specimens on weight average particle size (white column) and surface volume particle size (slash column) measured at 25°C.

3.3. Adsorption property

As shown in Figs. 5, 6, and 7, the absorption of MB increased with time. The decolorization of sample 1 (SA_2MWCNT_0) reached 40.16% after 120 h, meaning that SA itself has the ability to absorb MB. Compared with sample #1, the decolorizations of 2#, 3#, 4#, 5#, and 6# ordered by increasing content of MWCNT, were 55.78, 66.62, 76.9, 82.06, and 83.46%, respectively, when tested under the same conditions. This trend of increased decolorization with increased MWCNT concentration is attributed to the fact that the surface of MWCNT has substantial amounts of carbonyl that reacted with MB (see Fig. 8). Another reason for this decolorization may be due to the large specific surface area of MWCNT that greatly affects adsorption ability. Voids present in the MWCNT may also favor of the adsorption of MB.

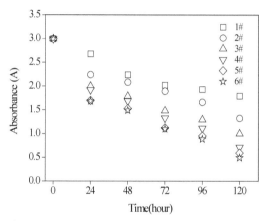

Figure 5. Absorbance of MB by SA/MWCNT microsphere, determined at 25°C.

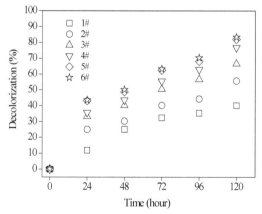

Figure 6. Decolorization of MB by SA/MWCNT microsphere determined at 25°C.

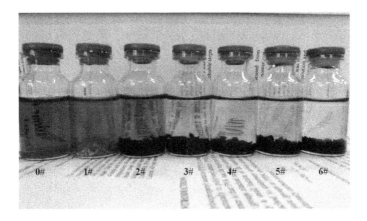

Figure 7. The photograph of MB absorbtion by different amounts of MWCNT.

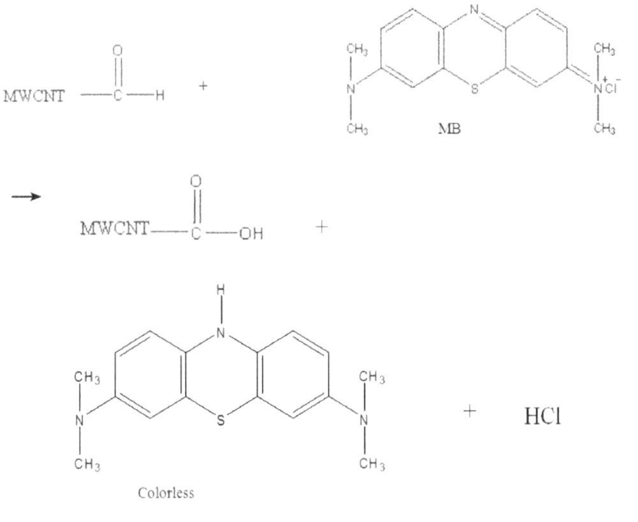

Figure 8. Reaction of MB by SA/MWCNT microsphere determined at 25°C.

The pH values decreased appreciably in samples 5# and 6# over the course of 120 h. The reason for this decrease may be the same as the reasons for decolorization previously mentioned (see Fig. 9), but the pH value of the original sample (50 mL of MB solution) was virtually unchanged. The reaction generated much more HCl that decreased the pH values, but the reaction rate eventually diminished after 120 h because there was less HCl generated and the adsorption of MWCNT surfaces was also nearly complete.

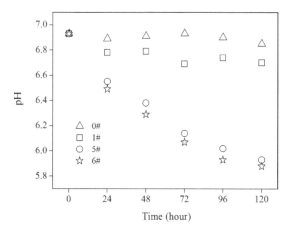

Figure 9. pH values of MB by SA/MWCNT microsphere determined at 25°C.

3.4. Electrical conductivity

The electrical conductivity was initially fixed at 79.3 μS/cm for the original sample, but increased considerably with increasing reaction time. The electrical conductivity of samples 5#, 6# increased sharply over the course of 120 h, but the electrical conductivity of the original sample (50 mL of MB solution) remained virtually unchanged, indicating that the original sample was stable.

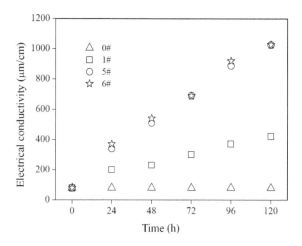

Figure 10. Electrical conductivity of MB by SA/MWCNT microsphere determined at 25°C.

4. Conclusions

This study effectively analyzed the adsorption of MB using gel beads prepared by sol–gel with SA and MWCNT. The formation conditions and mechanism of adsorption of the gel beads were also discussed. The decolorization of MB showed that the stability and reusability of SA/MWCNT could prove potentially advantageous in wastewater treatment.

Author details

Fang-Chang Tsai[1*], Ning Ma[1], Lung-Chang Tsai[2], Chi-Min Shu[2], Tao Jiang[1], Hung-Chen Chang[3], Sheng Wen[4], Chi Zhang[1], Tai-Chin Chiang[2], Yung-Chuan Chu[2], Wei-Ting Chen[2], Shih-Hsin Chen[5], Han-Wen Xiao[1], Yao-Chi Shu[6] and Gang Chang[1]

*Address all correspondence to: tfc0323@gmail.com

1 Ministry of Education Key Laboratory for the Green Preparation and Application of Functional Materials, Faculty of Materials Science and Engineering, Hubei University, P. R. China

2 Department of Safety, Health, and Environmental Engineering, National Yunlin University of Science and Technology, Douliou, Yunlin, Taiwan ROC

3 Department of Chemical and Materials Engineering, National Chin-yi University of Technology, Taichung, Taiwan ROC

4 Faculty of Chemistry and Materials Science, Hubei Engineering University, P. R. China

5 Department of Food Science, National I-Lan University, I-Lan, Taiwan ROC

6 Department of Cosmetic Applications & Management, Lee Ming Institute of Technology, Taipei, Taiwan ROC

References

[1] Ago, H., Kugler, T., Cacialli, F., Salaneck, W. R., Shaffer, M. S. P., & Windle, A. H. (1999). *J Phys Chem B*, 103, 8116-8121.

[2] Ajayan, P. M., Ebbesen, T. W., Ichihashi, T., Iijima, S., Tanagaki, K., & Hiura, H. (1993). *Nature*, 362, 522-523.

[3] Allen, S. J., & Koumanova, B. (2005). *J Univ Chem Technol Metall*, 40(3), 175-192.

[4] Banerjee, S., Hemraj-Benny, T., & Wong, S. S. (2005). *Adv. Mater*, 17, 17-29.

[5] Baughman, R. H., Zakhidov, A. A., & de Heer, W. A. (2002). *Science*, 297, 787-792.

[6] Braccini, I., & Perez, S. (2001). *Biomacromolecules*, 2, 1089-1096.

[7] Braccini, I., Grasso, R. P., & Perez, S. (1999). *Carbohydr Res*, 317, 199-130.

[8] Chen, C., Liang, B., Lu, D., Ogino, A., Wang, X., & Nagatsu, M. (2010). *Carbon*, 48, 939-948.

[9] Chen, C., Liang, B., Ogino, A., Wang, X., & Nagatsu, M. J. (2009). *J Phys Chem C*, 113, 7659-7665.

[10] Chen, J., Hamon, M. A., Chen, H. H. Y., Rao, A. M., Eklund, P. C., & Haddon, R. C. (1998). *Science*, 282, 95-98.

[11] Chen, K. C., Wu, J. Y., Huang, C. C., Liang, Y. M., & Hwang, S. C. J. J. (2003). *Biotechnol*, 101, 241-252.

[12] Courtois, J., Courtois, B., Heyraud, A., Colin-Morel, P., Dantas, L., Stadler, T., & David, P. (1993). *Agro-Food-Industry Hi- Tech.*, 6, 31-34.

[13] Ghosh, D., & Bhattacharyya, K. G. (2002). *Appl Clay Sci.*, 20, 295-300.

[14] Hameed, B. H., & Ahmad, A. A. (2009). *J. Hazard. Mater.*, 164, 870-875.

[15] Ho, Y. S., Chiang, T. H., & Hsueh, Y. M. (2005). *Proc Biochem*, 40(1), 119-124.

[16] Iijima, S. (1991). *Nature*, 354, 56-58.

[17] Ledakowicz, S., Solecka, M., & Zylla, R. J. (2001). *Biotechnol.*, 89, 175-184.

[18] Liu, J., Rinzler, A. G., Dai, H., Hafner, J. H., Bradley, R. K., & Boul, P. J. (1998). *Science*, 280, 1253-1256.

[19] Mawhinney, D. B., Naumenko, V., Kuznetsova, A., Yates, J. T., Jr., Liu, J., & Smalley, R. E. (2000). *J Am Chem Soc*, 122, 2383-2384.

[20] Mittal, A. K., & Gupta, S. K. (1996). *Water Sci Technol*, 34(10), 81-87.

[21] Moe, S. T., Draget, K. I., Skja, K. B. G., & Smidsrod, O. (1995). Marcel Dekker, New York, USA, 245-286.

[22] Morris, E. R., Rees, D. A., Thom, D., & Boyd, J. (1978). *Carbohydr Res*, 66, 145-154.

[23] Murakamia, M., Satob, N., Anegawac, A., Nakadad, N., Haradad, A., Komatsue, T., Takadaa, H., Tanakaf, H., Onoc, Y., & Furumaig, H. (2008). *Water Research*, 42, 2745-2755.

[24] Olson, T. M. (1994). *Water Research*, 6, 1383-1391.

[25] Pe'rez, S., Kouwijzer, M., Mazeau, K., & Engelsen, S. B. J. (1996). *Mol. Graph.*, 14, 307-321.

[26] Pierce, J. (1994). *J Soc Dyers Color*, 110, 131-134.

[27] Ray, S. S., & Okamoto, M. (2003). *Prog Polym Sci*, 28, 1539-1641.

[28] Ray, S. S., Yamada, K., Okamoto, M., & Ueda, K. (2002). *Polymer*, 44, 857-866.

[29] Renmin, G., Mei, L., Chao, Y., Yingzhi, S., & Jian, C. J. (2005). *Hazard. Mater*, 121, 247-250.

[30] Sahoo, N. G., Rana, S., Cho, J. W., Li, L., & Chan, S. H. (2010). *Prog Polym Sci*, 35, 837-867.

[31] Seshadri, S., Bishop, P. L., & Agha, A. M. (1994). *Waste Manage*, 15, 127-137.

[32] Simmons, J. M., Nichols, B. M., Baker, S. E., & Marcus, M. S. (2006). *J Phys Chem B*, 110, 7113-7118.

[33] Thakur, B. R., Singh, R. K., & Handa, A. K. (1997). *Crit. Rev. Food Sci. Nutr.*, 37, 47-73.

[34] Thom, D., Grant, G. T., Morris, E. R., & Rees, D. A. (1982). *Carbohydr Res*, 100, 29-42.

[35] Thostenson, E. T., Ren, Z., & Chou, T. W. (2001). *Compos. Sci.Technol*, 61, 1899-1912.

[36] Tsai, F. C., Shu, C. M., Tsai, L. C., Ma, N., Wen, Y., & Wen, S. (2011). Carbon nanotube industrial applications. In: Jose Mauricio Marulanda (Editor), *Carbon Nanotubes/ Book 2*, Rijeka, InTech-Open Access Publisher, 387-404.

[37] Tseng, C. H., Wang, C. C., & Chen, C. Y. (2006). *Nanotechnology*, 17, 5602-5606.

[38] Yasemin, B., & Haluk, A. (2006). *Desalination*, 194, 259-267.

[39] Zschoerper, N. P., Katzenmaier, V., Vohrer, U., Haupt, M., Oehr, C., & Hirth, T. (2009). *Carbon*, 47, 2174-2185.

Chapter 10

Carbon Nanotubes for Energy Applications

Dennis Antiohos, Mark Romano, Jun Chen and
Joselito M. Razal

Additional information is available at the end of the chapter

1. Introduction

1.1. The energy problem

The energy crisis during the 1970s sparked the development of renewable energy sources and energy conservation measures. As supply eventually met demand, these programs were scaled back. Ten years later, the hazards of pollution led to work on minimisation and reversal of the environmental impact of fossil fuel extraction, transport and consumption [1]. The United States Department of Energy predicts that 20 years from now, the world's energy consumption will increase by 20% (Figure 1). The growing concerns over the constant use of fossil fuels and its effect on climate change [2], has once again spurred research on sustainable energy development and on enhancement in renewable energy systems. Advances in energy storage and conversion systems that will make our energy usage more efficient are essential if we are to meet the challenge of global warming and the finite nature of fossil fuels [2, 3].

The need for the development of efficient energy storage systems is paramount in meeting the world's future energy targets, especially when energy costs are on the increase and more people need access to electricity [4, 5]. Energy storage technologies can improve efficiencies in supply systems by storing the energy when it is in excess, and then release it at a time of high demand [4]. Further material progression in research and development fundamentals, as well as engineering improvements need to be continued in order to create energy storage systems that will help alleviate humanities energy storage and conversion dilemmas.

Low grade heat (around 130°C) is a by-product of almost all human activity, especially when energy conversion is involved. It is also known as "waste heat" because the dissipated heat into the environment is unutilised. Progress in the field of thermal energy conversion can lead to effective use of limited fossil fuels and provide supplemental power to current energy conversion systems [6].

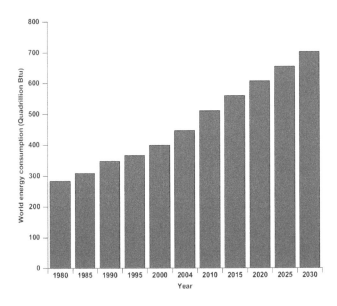

Figure 1. The United States Department of Energy values and forecasts for energy utilisation in the period from 1980 to 2030 [5]. Reproduced with permission from Elsevier.

1.2. How and why carbon nanotubes can address the issues of energy storage and conversion

Nanostructured materials are of great interest in the energy storage and conversion field due to their favourable mechanical, and electrical properties [3, 7]. Carbon nanotubes (CNTs) are one type of nanostructured material that possess these favourable electrical and mechanical properties due to the confinement of one dimension, combined with the surface properties that contribute to the enhanced overall behaviour. The potential of nanostructured materials is not only limited to energy storage and conversion devices; but also to nanotransistors [8, 9], actuators [8, 9], electron field emission [8, 9], and biological sensing devices [10, 11].

The use of carbon-based nanomaterials as electrode materials is practical and economically viable because cheap carbon pre-cursor materials are abundant [12]. As the research into carbon nanotubes (CNTs) has increased over the last 20 years, the cost of these materials has significantly reduced alongside improvements in processability and scalability [13].

The advantage of incorporating carbon materials and specifically CNTs as part of the electrode material is the excellent mechanical and electrical properties. They provide mechanical support to the substrate while enhancing the conductive and electrochemical properties. The low cost of the carbon precursor material used to synthesise CNTs makes device fabrication scalable and economically viable [14]. CNT assemblies can have extremely high specific surface areas, which are extremely important in capacitor design. CNT electrode materials can be confined to a smaller area increasing the electrode-electrolyte contact and decreasing the weight of the device therefore maximising the overall gravimetric performance of the device [15]. CNTs are also chemically stable, which enhances the resistance to degradation of the electrode surface [16].

2. Supercapacitors

2.1. Background information

Electrical energy can be stored in two different forms and can best be described when considering a battery and a capacitor. In a battery, it is the available chemical energy through the release of charges that performs work when two electroactive species undergo oxidation and reduction [17]; this is termed a Faradaic reaction. In a capacitor, electrostatic forces between two oppositely charged plates will separate charge. The generated potential is due to an excess and deficiency of electron charges between the two plates without charge transfer taking place [17]. The current that is observed can be considered as a displacement current due to the rearrangement of charges [2]; this effect is termed as non-Faradaic in nature.

2.1.1. Supercapacitor operation and types

There are two types of electrochemical capacitors that are referred to as 1) electric double layer capacitors (EDLC) and 2) pseudocapacitors. The construction of these devices can vary, with electrodes being fabricated from porous carbon materials including activated carbons, graphene, carbon nanotubes, templated carbons, metal oxides and conducting polymers [18, 19]. EDLC or supercapacitors have two electrodes immersed in an electrolyte solution, separated by a semi-permeable dielectric that allows the movement of ions to complete the circuit but prevents a short circuit from being formed. EDLCs are advantageous as they are able to provide relatively large power densities and larger energy densities than conventional capacitors, and long life cycles compared to that of a battery and ordinary capacitor [20]. The performance of supercapacitors is affected by the power density requirements, high electrochemical stability, fast charge/discharge phenomena, and low self-discharging [21]. Table 1 below shows a comparison between the three types of devices.

Parameters	Capacitor	Supercapacitor	Battery
Charge time	$10^{-6} - 10^{-3}$ sec	1- 30 sec	0.3 – 3 hrs
Discharge time	$10^{-6} - 10^{-3}$ sec	1 – 30 sec	1 – 5 hrs
Energy Density (Wh/kg)	<0.1	1 - 10	20 - 100
Power Density (W/kg)	>10 000	1000 - 2000	50 - 200
Cycle life	>500 000	> 100 000	500 - 2000
Charge/discharge efficiency.	≈ 1	0.90 – 0.95	0.7 – 0.85

Table 1. Comparison of key parameters for a capacitor, supercapacitor and battery [22].

Energy storage is achieved by the build-up and separation of electrical charge that is accumulated on two oppositely charged electrodes as shown in Figure 2 [12]. As stated previously, no charge transfer takes place across the electrode-electrolyte interface and the current that is measured is due to a rearrangement of charges. The electrons involved in the non-Faradaic electrical double layer charging are the conduction band electrons of the electrode. These electrons leave or enter the conduction band state depending on the energy of the least tightly bound electrons or the Fermi level of the system [2]. Supercapacitors exhibit very high energy storage efficiencies exceeding 95 % and are relatively stable for up to10^4-10^5 cycles [4, 5]. The energy given by the equation, $E = 0.5CV^2$, means that the operating voltage is the key in determining the energy characteristics of a supercapacitor. The choice of electrolyte when designing and fabricating a supercapacitor device dictates the operating voltage [23]. Operating voltages are approximately 1.2V, 2.7V, and 3.5V respectively for aqueous, organic and ionic liquid with all of them having associated advantages and disadvantages [4, 5].

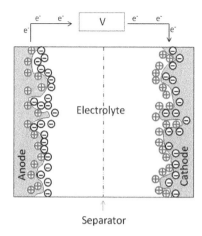

Figure 2. Schematic diagram of an EDLC supercapacitor with a positive and negative electrode, separator and porous carbon.

Like EDLC, a pseudocapacitor consists of two porous electrodes with a separator between them all immersed in an electrolyte solution [24]. However, the difference is that the charge is accumulated during Faradaic reactions near to or at the surface of the electrodes [18], hence non-Faradaic double-layer charging and Faradaic surface processes occur simultaneously [25]. The pseudocapacitance arises from a Faradaic reaction when some of the charge (q) passed in an electrode process is related to the electrode potential (V) via thermodynamical considerations [26]. The two principal cases are adsorption pseudocapacitance arising in underpotential deposition processes [26], and homogeneous redox pseudocapacitance where the reaction is reversible [26, 27]. Pseudocapacitors thus combine features of both capacitors and batteries [18, 28]. A comparison of energy density and power density for various electrical energy storage systems is depicted in Figure 3. Current commercial uses of supercapacitors include personal electronics, mobile telecommunications, back-up power storage, and industrial power and energy management [29, 30]. A recent application is the use of supercapacitors in emergency doors on the Airbus A380, highlighting their safe and reliable performance [30].

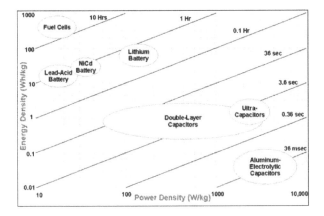

Figure 3. Ragone plot showing specific power against specific energy for various electrical energy storage systems. The times shown are the time constants of the device, which are obtained by dividing the energy density and power density [31].

2.2. Different electrolytes used and their advantages and disadvantages

The choice in electrolyte is extremely important for supercapacitor design as it influences the performance for energy storage and delivery. The extremely large surface area can allow for enhanced energy and power density as long as the micro-porosity and meso-porosity is tailored to suit the type of electrolyte being used. The electrolyte can be designed to enhance the cyclability, to sustain target power densities during operation, and to have an excellent rate capability (i.e. excellent charge/ discharge behaviour) [32].

2.2.1. Aqueous electrolytes

For aqueous electrolytes, the maximum operating voltage is theoretically limited by the reduction potential of water (1.23 V at room temperature) [32]. Most aqueous electrolyte systems tend to have an electrochemical window of approximately 1 V [32]. Electrolyte conductivity has a significant effect on the equivalent series resistance (esr) of the cell, which determines the power output [4]. Concentrated electrolytes are required to minimise the esr and maximise power capability [32]. In general, strong solutions of acidic electrolytes are much more corrosive than strong basic electrolytes meaning that the electrolyte has to be carefully selected for the particular electrode material.

Aqueous electrolytes tend to have very good kinetic behaviour of the electrolyte ions leading to very efficient charge/discharge rates. This behaviour is due to the relatively high conductivity and low viscosity of the concentrated solutions [4, 33]. For example, the conductivity of 1M H_2SO_4 is 730 mS cm^{-1} compared to the much lower value of 10-20 mS cm^{-1} for organic solutions of lithium salts [34]. Time constants of symmetrical carbon supercapacitors using H_2SO_4 were reported to be 0.1s [34].

2.2.2. Organic electrolytes

The use of non-aqueous electrolytes in supercapacitors has the main advantage of higher operating voltages compared to aqueous systems. Voltage windows can range up to 2.5 V and since the stored energy increases as V^2, it is possible to attain large energy and power densities [32, 34]. It must be noted that to operate at these higher voltages, non-aqueous electrolytes must be free of water and oxygen which will ensure no evolution of O_2 and/or H_2O at potential differences above 1.23 V [35]. Salts are added to the system to provide mobile ion movement at the electrode/electrolyte interface. The most common salt used generally consists of lithium ions as these ions move very well under an electric field and the effective ion diameter is very small [17].

The major disadvantages of non-aqueous systems are the lower conductivity and the higher viscosity resulting in higher equivalent series resistance (ESR) and reduced wettability if the electrode is hydrophilic. A decrease in wettability will effectively reduce the surface area used by the electrolyte, reducing the energy and power density. Most commercial systems that use organic electrolytes are manufactured in inert atmospheres and are costly to be produced [4].

2.2.3. Ionic liquids

Ionic liquids are another class of electrolytes that is proving a great area of research for electrolytes in supercapacitors. These electrolytes can be considered as molten salts with melting temperatures usually below room temperature where the ionic conductivity is no more than 20 mS cm^{-1} [34].

Common ions include BF_4^-, PF_6^- $(CF_3SO_2)_2N^-$, $CF_3SO_3^-$ as well as imidazolium, pyridinium, and quaternary ammonium salts [36]. The physical properties depend on the type of anion and cation and the alkyl chain length [37]. The main advantages are the good solvating properties, relatively high conductivity, non-volatility, low toxicity, large potential window, negligible vapour pressure and good electrochemical stability [37, 38]. Disadvantages include high viscosities and low conductivities compared to that of aqueous electrolytes; while some ionic liquid mixtures yield a potential window that is not much greater than that of aqueous systems. Capacitances approaching 100 F/g for activated carbon (AC) electrodes have been reported by Frackowiak et. al. by using $(CF_3SO_2)_2N^-$ anions and phosphoniumcations. Balducciet et. al. reported capacitance values of 115 F/g for asymmetric poly(3-methyl-thiophene)/AC electrodes using 1-buytl-3-methyl-imidazolium ionic liquids [36, 37].

2.3. Carbon nanotube powders

Carbon nanotubes (CNTs) were first discovered in 1953 through research in the Soviet Union, but the first accessible results were by Sumio Iijima [39], in 1991 as a result of research into buckminster fullerenes. CNTs have a cylindrical shape that can be considered as a graphene sheet rolled up; either individually as a single-walled carbon nanotube (SWNT), or concentrically as a multi-walled carbon nanotube (MWNT) as depicted in Figure 4and Figure 5. However, these sheets can have varying degrees of twist along its length that can lead to the nanotubes to be either metallic or semi-conducting as the change in chiralities induces different orbital overlaps [9]. They exhibit remarkable electrical transport and mechanical properties [7], which is why interest and research into this material has increased over the last two decades. CNT powders have the potential to be tailored to specific energy storage and conversion applications with there being an added advantage that they can be used in all electrolyte environments that encompass aqueous, organic and ionic liquids [40].

2.3.1. CNT synthesis overview

There are a variety of different methods for making SWNTs and MWNTs that have been developed since CNTs were first discovered. These include laser ablation, arc discharge, chemical vapour deposition (CVD), and high pressure carbon monoxide disproportionation (HiPCO). Recently, work by Harris et. al. has successfully scaled-up the synthesis of CNT using a fluidised bed reactor [41]. All growing conditions for synthesising CNTs require a catalyst to achieve high yields, where the size of the catalyst nanoparticles will determine the diameter and chirality of the CNT [42]. The CNTs that are formed are generally in a mixture with other carbonaceous product including amorphous carbon and graphitic nanoparticles.

2.3.2. Main synthesis methods for CNT growth

Both Laser ablation and arc-discharge methods for the growth of CNTs involve the condensation of carbon atoms generated from the evaporation of carbon sources. High temperature is involved, ranging from 3000 °C - 4000° C [43]. In arc discharge, various gases such as Helium or Hydrogen are induced into plasma by large currents generated at a carbon anode and cathode. This process leads to the evaporation of carbon atoms which produces very high quality MWNTs and SWNTs [44, 45]. Laser ablation also produces very high quality CNTs with a high degree of graphitisation by focusing a CO_2 laser (in pulsed or in continuous wave mode) for a period of time onto a rotating carbon target [46]. The HiPCO process utilises clusters of Fe particles as catalysts to create very high quality SWNTs [47]. Catalyst is formed *in situ* by thermal decomposition of iron pentacarbonyl, which is delivered intact within a cold CO flow and then rapidly mixed with hot CO in the reaction zone. Upon heating, the $Fe(CO)_5$ decomposes into atoms that condense into larger clusters. SWNTs nucleate and grow on these particles in the gas phase [48, 49].

The CVD method usually consists of a furnace, catalyst material, carbon source, a carrier gas, a conditioning gas, and a collection device (usually a substrate). The carrier gas is responsible for taking the reacting material onto the substrate where CNT growth occurs at catalyst sites [43]. The components mentioned are essential; however, different groups and researchers have alternative experimental conditions which can contain multiple types of furnaces, and a variety of catalyst and carbon sources. The key advantage of this technique is its capability to directly deposit the CNTs onto the substrate, unlike arc discharge and laser ablation that produces a soot / powder. Recent developments by Harris et. al. [41] has led to the development of a large scale batch process for fabricating MWNTs. Here, a furnace like system called a fluidised bed reactor continuously flows a carrier gas over a porous alumina powder that is impregnated with the catalyst material, leading a to continuous creation of MWNTs where tens of grams can be synthesised in one run.

2.3.3. Single walled nanotubes

SWNTs have been studied extensively as a supercapacitor and hybrid energy material [4, 50, 51]. The structure of a SWNTis illustrated in Figure 4 with a cylindrical nature apparent as previously stated. Its advantage is that it has very good thermal and conductive properties where the thermal conductivity can exceed 6000 $Wm^{-1}K^{-1}$ and a potential current carrying capacity of 10^9 A/cm^2 [52, 53].

The maximum reported gravimetric capacitance for SWNT fabricated electrodes (PVA / PVC binder; pressed into pellet) is 180 F/g with an energy density of 7 Wh/kg and a power density of 20 kW/kg using KOH electrolyte [54, 55]. Hu et. al. [56] have recently reported a solid state paper based SWNT supercapacitor, which has a specific capacitance of 115 F/g, energy density of 48 Wh/kg and a large operating voltage of 3V. The electrode preparation involved pre-processing where cotton sheets were immersed in the SWNT dispersion, annealed then immersed in an PVA/ H_3PO_4 electrolyte.

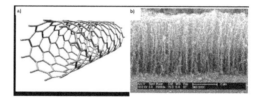

Figure 4. (a) Schematic representation of a SWNT[57].(b) FESEM of SWNTs grown onto a Si wafer substrate[58]. Reproduced with permission from Elsevier.

2.3.4. Multi-walled nanotubes

Like their SWNT counter parts, MWNTs have also been studied extensively as electrode materials for supercapacitors [4, 50, 51]. The advantages over SWNTs are their ability to be more easily synthesised on much larger scales, making them more suitable for commercial application. The concentric nature of MWNTs can be observed in the SEM image of Figure 5. The maximum gravimetric capacitance attained for electrodes constructed from MWNTs range between 4-140 F/g with the best available commercial result at 130 F/g from Maxwell's Boost capacitor [59]. Wang et. al. [50] have recently reported partially exfoliated MWNTs on carbon cloth that gave a specific capacitance in the range of 130-165 F/g with a coulombic efficiency of 98 %.

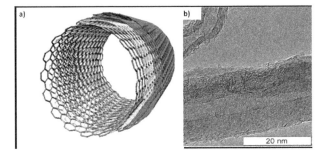

Figure 5. (a) Schematic representation of a SWNT [57]. (b) TEM images of pristine MWNTs [59]. Reproduced with permission from Elsevier

2.3.5. Surface functionalities

The presence of surface functionalities such as oxygen, nitrogen, hydrogen, boron and catalyst nanoparticles (dependent on the synthesis environment and pre-cursor materials) can affect the capacitative behaviour of the electrode through the introduction of Faradaic reactions [60], changes in electric and ionic conductivity [23], and influencing wettability [61]. A schematic representation of an sp^2 hybridised carbon lattice with various dopants is show in Figure 6.

Oxygen

Carbon materials will have functional groups present on their surface as a result of the precursors and preparation conditions [23]. Most of these functional groups are in the form of –COOH, =CO as well as phenol, quinone and lactone groups [4, 23]. Activation procedures such as post treatment with H_2SO_4 and / or HNO_3 also leads to acid oxygen functionalities [4].

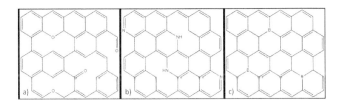

Figure 6. Schematic representation of a sp^2 hybridised carbon lattice that has been doped with; (a) oxygen functional groups, (b) nitrogen functional groups, (c) boron.

Most of these groups are bonded with carbon atoms at the edge of hexagonal carbon layers where Faradaic reactions (via interactions with the electrolyte) lead to pseudocapacitance such as those developed with transition metal oxides RuO_2 and MnO_2 [23, 62]. These functional groups can also be purposely added onto the surface of carbons via oxidation with O_2 or acid treatment with HNO_3 or H_2SO_4 [63]. In aqueous systems, the presence of oxygen containing functional groups can lead to an enhanced wettability as well as pseudocapacitance as mentioned above, which maximises the electroactive surface area leading to larger energy densities [8, 23]. It has been proposed that the pseudocapacitative reactions for oxygen functionalised CNTsinvolve carboxyl groups undergoing electron transfer[64]:

$$C-OH \; -- \; C=O + H^+ + e^-$$

$$C=O + e^- \; -- \; C - O^-$$

In non-aqueous systems, however, oxygen functionalities are detrimental to device performance. Parasitic redox reactions can lead to a degradation of the electrode, as well as adverse effects relating to voltage proofing and increased leakage current [4, 65]. These redox reactions will reduce the cycle life of a device, as well as lowering the operating voltage.

Shenet. al. [66] reported in 2011 the effects of changing the carboxylic group concentration on SWNTs. The specific capacitance, power density and energy density 0.5 M H_2SO_4 electrolyte increased with carboxylic group density reaching a maximum of 149.1 F/g, 304.8 kW/kg, and 20.71 Wh/kg, respectively. The 10 μm film electrodes were fabricated using vacuum filtration to create "bucky papers" onto a mixed cellulose estate membrane.

Nitrogen

Nitrogen containing carbons have recently attracted interest due to its n-type behaviour that promotes large pseudocapacitance, which can be obtained even if the surface area of the material is decreased [67]. In some instances, up to 3-fold increase in capacitance have been reported [68]. Typical examples of redox reactions involving nitrogen are described below [69]:

$$C + NH + 2e^- + 2H^+ \leftrightarrow CH_2-NH_2$$

$$CH-NHOH + 2e^- + 2H^+ \leftrightarrow CH_2-NH_2 + H_2O$$

The chosen precursor material affects the types of functionalities that are attached to the carbon backbone. Nitrogen-containing groups may be added via various methods with compounds containing nitrogen including treatment with urea, melamine, aldehyde resins and polyacylonitrile [4, 70-73]. Surface areas for nitrogen-doped carbon materials are thought to be in excess of 400 m^2/g [23]. This is much lower than pure SWNTs and pure MWNTs that have been reported to attain a surface area greater than 1315 m^2/g and 830 m^2/g respectively; suggesting that pseudocapacitance makes up a substantial portion of the total capacitance [74, 75]. Y. Zhang et. al. [76] have showed that N-doped MWNTs synthesised via CVD growth exhibited a capacitance of 44.3 F/g, which was more than twice the value obtained than that of the un-doped MWNTs in a 6M KOH electrolyte. K. Lee et. al. [77] have shown that the nitrogen content on vertically aligned CNTs increases the capacitance until a certain point due to an increased donation of an electron by the N (N acts as an n-type dopant) and an enhanced wettability in aqueous systems. Excessive N-doping significantly reduced the conductivity and inhibited charge storage and delivery [77]. The doped and un-doped CNTs were directly grown onto a stainless steel substrate using CVD [77].

Boron

Boron is another interesting material for doping CNTs due to its p-type nature which promotes CNT growth and increases the oxidation temperature of the nanotubes [78]. However, the development of boron doped CNTs for the use as electrodes in supercapacitor devices is not well established [23]. Work by Shiraishi et.al. showed that boron doping MWNTs, increased the capacitance per surface area from 6.5 $\mu F/cm^2$ to 6.8 $\mu F/cm^2$ in 0.5M LiBF$_4$/PC [79]. These electrodes were once again synthesised using CVD [79]. Wang et. al. reported in 2008 that interfacial capacitance was increased by 1.5-1.6 times in boron-doped carbon than that in boron-free carbon with alkaline electrolyte (6 M KOH) and/or acid electrolyte (1 M H2SO4) [80]. The carbon material was made into a slurry using carbon black and PTFE binder and pasted onto a Ni mesh current collector [80].

2.3.6. Advantages, limitations and comparison

It can be seen that CNTs can be tailored different ways in order to tune (to a degree) the performance of the electrode material. This control has been demonstrated by firstly, varying the chiralitiy of the nanotube to produce the single-walled or the multi-walled variety. Both CNT types have associated advantages and disadvantages with SWNTs being able to be synthesised with a high degree of purity; while MWNTs can be synthesised on a larger scale. CNTs can also have functionalities (through addition of oxygen or nitrogen containing groups) added to their structure through treatment in order to change the surface properties and hence wettability of the material. These functionalities enable enhanced compatibility to an electrolyte to maximise electroactive surface area usage and hence performance. Further doping with specific elements such boron and nitrogen can introduce a p-type/n-type behaviour where electrons contribute a Faradaic response to the system and enhance capacitance and energy density. However, it must be noted that when faradaic processes occur at the electrode/electrolyte interface, irreversible processes increase degradation of the electrode over time. Specific capacitance of CNTs (three electrode and device testing) is in the order of 5 – 165 F/g with an increase thereafter as a result of doping (i.e due to Faradaic contribution). It must be pointed out that with electrical energy devices, there is always a trade-off between energy and power density. Therefore the electrode material has to be tailored to meet the requirements of the specific application.

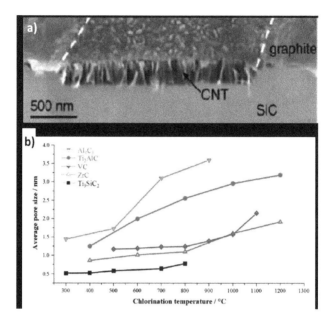

Figure 7. (a) SEM image depicting the growth of templated porous CNTs. (b) The tunability of average pore size distribution of binary and ternary carbides Al_4C_3, Ti_2AlC, VC, ZrC, Ti_3SiC_2 by varying the chlorination temperature [29]. Reproduced with permission from John Wiley & Sons

2.4. Templated porous carbons

Templated porous carbons are of recent great interest in the field of energy storage due to the tunability in porosity, which is necessary to meet the materials application requirements [29, 30]. These templated carbons are commonly known as carbide derived carbons (CDC) as the carbon materials are derived from carbon precursors through physical and/or chemical processes [29]. Briefly, the synthesis involves halogenations (usually chlorination) where the carbon is formed by selective extraction of the metal and metalloid atoms, transforming the carbide structure into pure carbon. The carbon layer is formed by inward growth, with retention of the original shape and volume of the precursor [29]. If any metal chlorides are trapped, they can be usually removed by hydrogenation or vacuum annealing [29]. The general reaction scheme is as follows where M = Si, Ti, Zr[23, 29];

$$MC(s) + 2Cl_2(g) \rightarrow MCl_4(g) + C(s)$$

The advantage of forming carbon structures this way is the ability to form a tailored and narrow pore size distribution with a large surface area as can be seen in Figure 7 [29].

Inagaki et. al. in their very comprehensive review of carbon materials for electrochemical capacitors reported a maximum surface area of S_{BET} of 2000 m²/g for CDC, which gives rise to possible electrode materials with extremely large energy densities and power densities [23].

Gao et. al. have recently reported flexible CDC electrodes fabricated into a device which obtained a specific capacitance of 135 F/g in 1 M H_2SO_4 and 120 F/g in 1.5 M tetraethyl ammonium tetrafluoroborate ($TEABF_4$) [81]. Ordered mesoporous carbon spheres with impregnated NiO, and a maximum surface area of 944 m^2/g yielded a specific capacitance of 205.3 F/g in 2 M KOH [82]. Reported also by Y. Korenblit [83] was a high surface area CDC (2430 m^2/g) with aligned mesopores, which yielded a specific capacitance of 170 F/g and an extremely high capacity retention of 85% at high current densities of 20 A/g.

2.5. Composite electrode materials

Typical carbonaceous electrode materials (activated carbon, CNTs, graphene, CDC) with high surface areas used in supercapacitors have somewhat reached a limit when it comes energy storage capacity, thus restricting their possible applications [84]. Pseudocapacitor materials that are able to meet the needs of higher energy density are currently being developed and combined with carbonaceous materials in order to create composites that when designed into hybrid supercapacitors have advantages of fast rate capability, high storage capacity, and long cyclability[84, 85].

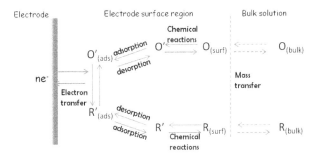

Figure 8. Schematic diagram of a reversible redox reaction, as well as EDLC occurring at the electrode/electrolyte interface leading to pseudocapacitance.

2.5.1. CNT / polymer

Electrode materials comprised of inherently conducting polymers (ICPs) and CNTs is a promising area of research. The conductive polymer matrix, combined with the network like structure of the CNTs provides an enhanced electronic and ionic conductivity that can considerably improve charge storage and delivery [86-88].

Antiohos et. al. reported a SWNT / Pedot-PSS composite electrode material that was fabricated into a device which had a specific capacitance of 120 F/g (1 M $NaNO_3$ / H_2O), coupled with an excellent stability (~90% capacity retention) over 1000 cycles using galvanostatic charge / discharge [89]. The SWNT / Pedot-PSS composite is depicted in Figure 9 where SWNTs are thoroughly dispersed throughout the Pedot-PSS conducting polymer matrix. Kim et. al. recently fabricated a ternary composite material consisting of MWNTs, graphene, and PANI where a specific capacitance of 1118 F/g was achieved. This electrode was stable with 85% capacity retention after 500 cycles using galvanostatic charge / discharge [90]. Hu et. al. [91] have recently reported a composite electrode materials containing MWNTs coated with polypyrrole that achieved a high capacitance of 587 F/g in a 0.1 M NaClO$_4$ / acetonitrile electrolyte.

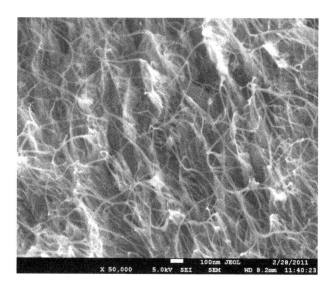

Figure 9. SEM image of PEDOT/PSS-SWNT composite showing PEDOT/PSS polymer to be integrated with the SWNT [89]. Reproduced with permission from The Royal Society of Chemistry.

2.5.2. CNT / metal oxide

Metal oxides exhibit pseudocapacitative behaviour over small rages of potentials, through redox processes which contribute electron transfer between the electrode / electrolyte interface (Figure 8). Common materials used in the construction of such devices are oxides of Mn, Ru, Ir, Pt, Rh, Pd, Au, Co and W [22, 26]. By combining metal oxides with CNTs, composites can be formed that combine both Faradaic and non-Faradic effects enabling a larger energy density to be obtained while still holding reasonable power density. Figure 10 shows MnO$_2$ particles that have been formed (*insitu*) in the presence of MWNTs.

Very recent work on carbon / metal oxide composites can be found in the review by Wang et. al. [50]. Myoungkiet. al. reported recently in their a RuO$_2$ / MWNT, electrode material which achieved a specific capacitance of 628 F/g[92]. The electrode was fabricated by dispersing the mixture in ethanol and casting onto carbon paper [92]. Li. et. al. reported that when MWNT were coated with MnO$_2$, a capacitance of 350 F/g was achieved [93]. More novel materials have been created by incorporating MWNTs and Co$_3$O$_4$which yielded specific capacitances of 200 F/g [94] (acetylene black / PVDF slurry on Ni gauze); while Jayalakshmi et. al. reported in 2007 V$_2$O$_5$.xH$_2$O / CNT film with a specific capacitance of 910 F/

gwith the material being ground into a paste with paraffin and spread onto a graphite electrode, and tested in 0.1 M KCl [95].

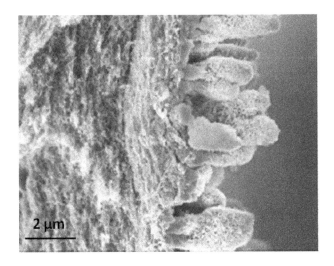

Figure 10. Surface cross-section morphology of MnO_2 particles being grown (insitu) onto MWNTs[96].Reproduced with permission from Elsevier.

2.5.3. CNT / carbons

Creating composite materials from CNT and different forms of carbon such as graphene or carbide derived carbons (CDC) can be advantageous due to the fact that the CNTs provide microporosity (large surface area to maximise capacitance and hence energy density); while graphene and CDC can be used to tailor the mesoporosity which improve ions kinetics, enhancing the power density [21]. In Figure 11, a composite of reduced graphene oxide coated with SWNTs is depicted that has been formed into a porous film. The edges of the graphene oxide protrude out with a uniform coating of SWNTs. Recently, Li et. al. fabricated different mass loadings of graphene and CNT composite electrodes by solution casting onto glass, annealing then peeling off [97]. They reported capacitance ranges of 70-110 F/g at a scan rate of 1 mV/s in 1M H_2SO_4 [97]. Luet. al. have reported a CNT / graphene composite which was bound together with polypyrrole (through a filtration process) that achieved a specific capacitance of 361 F/g at a current density of 0.2 A/g in 1 M KCl. The electrode exhibited excellent stability with only a 4% capacity loss over 2000 cycles [97]. Dong et. al. have shown that is it possible to form SWNT/graphene oxide core shell structures and spray coat the subsequent material onto a current collector [98]. The performance of these core structures yielded a material with a specific capacitance of 194 F/g using galvanostatic charge / discharge at a high current density of 0.8 A/g in 1 M KOH [98].

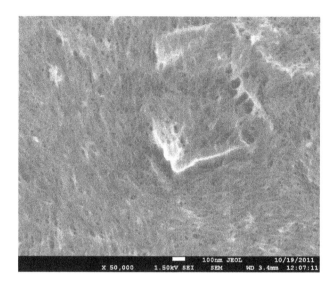

Figure 11. SEM image of a reduced graphene oxide / SWNT composite formed into a film.

2.6. Conclusions

It can be seen that there has been extensive research and development in the use of CNT as electrode materials for energy storage applications. Currently, they provide an excellent platform for devices that require high power density due to the very high surface areas and fast rate capability. Further studies need to be implemented in order to better understand the relationship between electrode porosity and electrolyte. An enhanced understanding of the role of micro and meso-porosity and its effect on system performance is critical. Electrolyte selection is also critical to device performance as it is proportional to the square of the voltage. The main classes of electrolytes are aqueous-based, organic-based and room temperature ionic liquids. Evolving work has focused on using CNT materials in conjunction with doping of various functional groups such as carboxyls, amines and elements such as boron and nitrogen in order to enhance the electrode performance through increased usage of electrode surface area and / or Faradaic contributions. The most recent work has focused on the creation of composite materials via the combination of CNTs with conducting polymers or metal oxides. CNT composites have amassed into a prevalent area of research through the search for the discovery of hybrid energy storage devices that are able to have high energy and high power density which are beneficial for creating more energy efficient systems and providing a greater range of applications.

3. Thermal Energy Harvesting

3.1. Introduction to thermogalvanic cells

Studies on the conversion of heat to electricity have been conducted as early as the 1960s [99]. Since then, several thermal converters have been developed: thermocouples, thermionic converters, thermally recharged cells, thermogalvanic cells, etc. [100]. The discussion in the subsequent sections will be limited to thermogalvanic cells, also known as thermocells. These are electrochemical systems that are able to directly transduce thermal energy to electrical energy [101]. The simple design of these systems allows them to function without the need for moving components. Their stability allows operation for extended periods without regular maintenance. Thermocells also have zero carbon emission hence it will not contribute to the environmental impact of electrical power generation.

3.1.1. Low grade heat sources and conversion through thermogalvanic cells

Various unharnessed low grade heat sources

The second law of thermodynamics dictates that a heat engine can never have perfect efficiency and will always produce surplus heat (usually around 100 °C). This waste heat (or low grade heat) is one of the world's most ubiquitous sources of untapped energy. (i.e. waste heat is produced by simply turning on an automobile). Roughly 70 % of the energy generated by an automobile motor is wasted; part of it ends up as a hot exhaust pipe and warm brakes. The Wartsila-Sulzer RTA96-C turbocharged two-stroke diesel engine, one of the most efficient engines in the world, is only able to convert around 50 % of the energy in the fuel to useful motion. The rest of that energy gets dissipated as waste heat [102]. Waste heat also exists in factories, particularly in the steel and glass production industry. Pipes that carry hot liquids are also low grade heat sources. Other scenarios wherein heat simply dissipates into the environment are power plants, household appliances, and various electronic gadgets. Research done to convert waste heat into electrical power by the use of ferromagnetic materials, thermocouples and thermionic converters has resulted in low efficiencies [103-105]. Advances in thermoelectric systems have been hampered by its high initial cost and material limitations; as these systems operate in the temperature range much higher than low grade heat [106].

Description of how a thermogalvanic cell works

A thermogalvanic cell, also known as a thermocell, is a thermal energy converter that utilises electrochemical reactions to attain conversion of low grade heat to electrical power. The two half cells of the system are held at different temperatures causing a difference in redox potentials of the mediator at the anode and cathode [107]. This reaction can drive electrons through an external circuit that allows generation of current and power. A schematic of a thermocell with a ferri/ferrocyanide redox couple is shown in Figure 12.

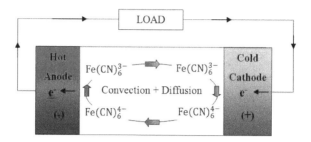

Figure 12. Ferri/Ferro Cyanide redox thermogalvanic cell [108]Reproduced with permission from Springer Science +Business Media

Ferrocyanide is oxidized at the hot anode, the electron generated then travels through an external circuit and returns to the cell via the cold cathode where it is consumed in the reduction of ferricyanide[109]. The accumulation of reaction products at either half-cell is prevented by the diffusion and convection of the electrolyte that occurs naturally, thus eliminating the need for moving mechanical components.

3.1.2. Desirable material properties for thermal conversion cells

Power conversion efficiency and how it is affected by electrode material properties

The power conversion efficieny (Φ) of a thermocell is defined as follows:

$$\Phi = \frac{\text{Electrical output power}}{\text{Thermal power flowing through the cell}} \qquad (1)$$

The thermal power flowing through the cell is largely controlled by cell design and electrolyte selection. When a reversible redox couple is used, no net consumption of the electrolyte occurs and the thermal power is given by:

$$\textbf{Thermal power flowing through cell} = \textbf{KA}\frac{\Delta T}{d} \qquad (2)$$

Where K is the thermal conductivity of the electrolyte, A is the electrode cross sectional area, ΔT is the thermal gradient and d is the distance between the two electrodes [110].

Qualitative behaviour of the current and voltage dependencies are shown in Figure 13; it depicts that the maximum electrical output power (P_{max}) is obtained when the external and internal loads are equal and is given by:

$$P_{max} = 0.25 V_{OC} I_{SC} \qquad (3)$$

Where V_{oc} is the open circuit voltage and I_{sc} is the short circuit current. V_{oc} is highly dependent on the reaction entropy of the redox couple and the thermal gradient at which the electrodes are exposed to as shown in Equation 4:

$$V_{oc} = \frac{\Delta S_{B,A} \Delta T}{nF} \tag{4}$$

where $\Delta S_{B,A}$ is the reaction entropy for a hypothetical redox couple $A \leftrightarrow ne^- B$, n is the number of reactions involved in the redox reaction and F is Faradays constant [100].

Combining Equation 2 and Equation 3 allows the power conversion efficiency to be expressed as:

$$\Phi = \frac{0.25 V_{oc} I_{sc}}{KA \left[\frac{\Delta T}{d} \right]} \tag{5}$$

Ohmic, mass transport and activation overpotentials are losses that need to be minimised in order to realise an improvement in thermocell conversion efficiency. At large electrode separations, ohmic overpotential is dictated by the electrolyte resistance; and mass transport overpotential is maximized. By decreasing the inter-electrode separation, an increase in generated power will be observed as both ohmic and mass transport overpotentials will decrease. However, the power conversion efficiency will be lowered as it will be harder to maintain the thermal gradient in the cell [112]. It has been shown that changes in electrolyte concentration affect its thermal conductivity [108]. Optimization of electrolyte concentration coupled with appropriate cell design is necessary to mitigate both overpotentials while maintaining large power conversion efficiency.

Activation overpotential is associated with the activation barrier needed to transfer an electrode to an analyte. For the same activation overpotential, larger current densities are realised when the exchange current density is increased. This increase is attained when the concentration of the redox couple in the electrolyte is maximised, the thermal gradient is increased and the number of possible reaction sites is augmented [113]. Porous electrodes have the advantage of increased electroactive surface area and will directly amplify the short circuit current density [114]. It must be noted that for porous electrodes, short circuit current density does not increase indefinitely with electrode thickness as mass transfer overpotential will become limiting. The reaction products formed within the pores of the anode will not be able to diffuse fast enough to the cathode and vice versa, generating concentration gradients around both electrodes. Another way to decrease the activation overpotential in thermocells is by using catalytic electrodes attained by doping [115].

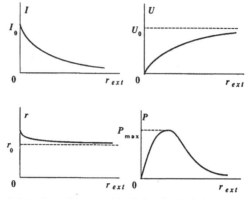

Fig. 2. The typical dependences of the current I, the effective voltage U, the internal resistance r and the useful power from the external resistance r_{ext}.

Figure 13. The typical dependencies of the current (I) on the effective voltage (U), internal resistance (r) and the useful power on the external resistance (r_{ext}) [111]Reprinted with permission from Elsevier.

3.1.3. CNTs vs flat electrodes - why CNTs can improve thermal harvesting

Power conversion efficiency achieved by using flat electrodes and why recent developments in CNTs can augment thermogalvanic cell performance

The chemical stability of platinum led to its extensive investigation in thermogalvanic cells. In fact, a study on the effects of platinum electrode cleaning was performed and it was deduced that this affects the power delivery characteristics of thermogalvanic cells [116].

Comparison of thermoelectric converters operating at different conditions (i.e. thermal gradient, electrolyte, electrode separation, etc.) can be done by measuring their power conversion efficiency relative to a Carnot engine operating at the same temperature (Φ_r).

$$\Phi_r = \frac{\Phi_{\text{thermogalvanic cell operating at } \Delta T}}{\Phi_{\text{Carnot engine operating at } \Delta T}} \qquad (6)$$

If power inputs are ignored, such as mechanical stirring, thermocells with platinum electrodes are able to attain power conversion efficiency relative to a Carnot engine of 1.2%. However, if power inputs are strictly excluded, the efficiency drops to 0.5% [100].

As mentioned previously, the discovery of CNTs led to widespread research on this material to investigate its potential uses, one of them being electrochemical applications [39, 117-120]. CNT electrodes are known to exhibit Nernstian behaviour and more importantly, fast electron transfer kinetics with the redox couple ferri/ferrocyanide. Peak potential separation in cyclic voltammograms obtained using micron-sized MWNT electrodes and 5 mM potassium ferrocyanide is 59 mV, which is the expected theoretical value and implies that the highest electron transfer rate was attained [121]. Incidentally, the ferri/ferrocyanide re-dox couple has been studied intensively in thermocell applications owing to the large volt-age that can be induced by a thermal gradient, also known as the Seebeck coefficient. The Seebeck coefficient 1.4 mV/K for the ferri/ferro cyanide redox couple implies that an open circuit potential of 84 mV is attainable at a thermal gradient of 60 °C (the usual limit for aqueous systems without significant cooling). The fast electron transfer of CNTs in ferri/ferro cyanide primarily justifies its use as electrodes in thermocells.

The nanometre diameter of CNTs gives rise to large gravimetric and volumetric specific sur-face areas (SSA). Their unique aspect ratios allow porous electrodes to be fabricated by a va-riety of methods. Theoretically the SSA of CNTs can range from 50-1315 m^2/g, the value dictated by the number of walls [74]. Theoretical predictions are in good agreement with ex-perimental values obtained by the measurement of amount of gas (usually N_2) adsorbed at 77 K and calculations using the Brunauer-Emmett-Teller (BET) isotherm. Kaneko et al. [122] have reported that MWNTs are mesoporous while Rao et al. [123] have shown that SWNTs are microporous. MWNT buckypapers of the same geometric area compared with platinum foil are known to have three times larger charging current density during cyclic voltamme-try in ferri/ferrocyanide aqueous electrolyte [114], evidence of the large accessible SSA of CNT electrodes.The large SSA of CNTs allows for a greater number of electroactive sites. When the CNT electrode porosity is controlled and the tortuousity is minimised in thermo-galvanic cells (so that mass transfer is not limited within the electrode), the short circuit cur-rent generated can be significantly augmented.

3.2. Different types of CNTs investigated

3.2.1. SWNT and MWNT

CNTs were first used as thermocell electrodes in 2009 [114]. Baughman et al. tested 0.5 cm^2 MWNT buckypaper electrodes (with less than 1 % catalyst and with MWNT diameter of around 10 nm) in a U-Cell with electrode separation of 5 cm, a temperature gradient of 60 °C wherein T_{cold}= 5 °C. A schematic of the cell they used is shown in Figure 14. A specific power density of 1.36 W/m^2 was obtained [114]. Platinum electrodes tested under the same condi-tions generated a specific power density of 1.02 W/m^2, proving that CNTs are viable materi-als for thermocell electrodes.

SWNT powders produced by arc discharge (ASA-100F, Hanwha Nanotech) with an average diameter of 1.3 nm, and composition of 20-30 wt. % CNTs, 40 wt. % carbon nanoparticles, 20 wt. % catalyst material, 10 wt% amorphous carbon and graphite, was tested by Kang et al. in thermal harvesting [113]. A vertical test cell with a "hot above cold" orientation (Figure 15), glass frit separator and electrode separation of 4 cm was employed with T_{hot}= 46.4 °C and T_{cold} = 26.4 °C. SWNT electrodes with an area of 0.25 cm^2 were immersed in 0.2M $K_3Fe(CN)_6$/$K_4Fe(CN)_6$ electrolyte. The specific power density obtained was around 5.15 W/kg. Commercially available purified SWNT powders (P-SWNT) sourced from Hanwha Nanotech (ASP-100F), refined by thermal and acid treatment (60-70 wt. % nanotubes, 10 wt. % catalyst material, 20 wt. % graphite impurities) was tested by the same group. The specific power density improved by 32 %, generating 6.8 W/kg. Using the same test conditions, commercially available purified MWNT having 3-6 walls with a median diameter of 6.6 nm (SMW100, Southwest Nanotechnologies, Inc) and approximately 98 wt. % carbon yielded a specific power density of 6.13 W/kg. It must be noted that these tests were not done to maximise the power generation capability of the thermocell but to gain further insight into CNT electrodes for thermal harvesting. Hence the small electrode separation, low electrolyte concentration and small thermal gradient.

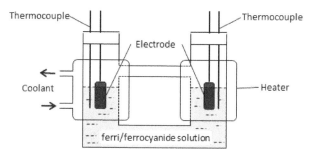

Figure 14. Schematic of the U-Cell used by Baughman et. al for thermal harvesting

Electrical impedance spectroscopy (EIS) of the various carbon nanomaterials tested by Kang et al revealed that the P-SWNT electrode has a marginally lesser ohmicoverpotential (21 Ω) than the pristine SWNT (22 Ω). This finding explains the increased specific power density generated when the P-SWNT electrodes are used.

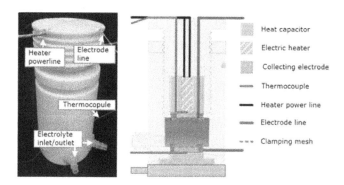

Figure 15. Vertical "hot above cold" thermocell[113]Reprinted with permission from John Wiley & Sons, Inc.

It has been proven that the catalytic nature of MWNTs is due to the edges or sites where the tube terminates, regions that are more numerous in MWNTs than in SWNTs [124]. Due to this and the fact that P-MWNTs had lower ohmic resistance (18 Ω) compared to P-SWNTs, it was expected that P-MWNTs would perform better in thermal harvesting. The authors attributed the enhanced performance of P-SWNT to the larger specific surface area, which compensated for the decreased electroactive sites and higher ohmic resistance.

3.2.2. Functionalized CNTs

Functionalising or doping (using Nitrogen or Boron atoms) may be used to fine tune the physical and chemical properties of CNTs [125-127]. With advances in technology, CNT functionalization is a reasonably simple process [128]. Dai et. al have shown that nitrogen-doped carbon nanotubes (NCNTs) have high electrocatalytic activity in oxygen reduction reactions as compared to undoped CNTs [129]. The increase in performance was brought about by a four electron pathway for oxygen reduction reactions that was attained by aligning the nanotubes and integrating nitrogen into the carbon lattice. The additional electrons contributed by nitrogen atoms can enhance electronic conductivity by providing electron carriers for the conduction band [130]. The many active defects and hydrophilic properties of NCNTs allow for enhanced electrolyte interaction in aqueous solutions [131]. Boron doped CNTs (BCNTs) are also attractive for electrochemical applications owing to the increased edge plane sites on the CNT surface; proven to be the predominant region for electron transfer [132]. Examples of the electrocatalytic performance of BCNTs are the improved detection of L-cysteine, enhanced electroanalysis of NADH and enhancement of field emission [126, 127,133].

The possibility of using nitrogen-doped CNT and boron-doped CNT electrodes in thermocells was investigated by Cola et.al [115]. Doping was attained by using a plasma-enhanced chemical vapour deposition process. Tests were run using a U-cell configuration (Figure 14), T_{cold} = 20 °C, thermal gradient of up to 40 °C, and electrolyte concentration of 0.1 M potassium ferri/ferrocyanide. The electrodes were sized to 0.178 cm^2 and were set up in a symmetric and asymmetric (N(hot)-B(cold) and B(hot)-N(cold)) fashion.

Results (Figure 16a) indicate inferior thermocell performance for both NCNT and BCNT as compared to Pt and pristine CNTs. The poor performance of the doped CNTs was brought about by the sluggish kinetics, evidenced by the large peak separations in the cyclic voltammograms taken at a scan rate of 100 mV/s (Figure 16b). It was theorized that the slow-moving kinetics for the doped electrodes was caused by the electrostatic effects at the electrode-electrolyte interface [134]. The positively charged BCNTs repulsed the similarly charged potassium (K$^+$) counter ion which decreased the electrolyte concentration in the vicinity of the electrode. The negatively charged NCNTs led to a strong electrostatic attraction with K$^+$, an effect which at low concentrations can improve electron transfer kinetics. However, the large bulk concentration (needed to achieve significant short circuit currents in thermocells) led to a high density of the K$^+$ ions in the vicinity of the electrode. This effectively acted as a barrier to the redox reactions that were supposed to occur at the electrode.

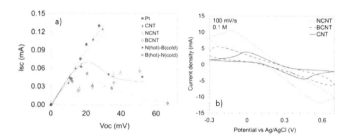

Figure 16. a) Thermal energy conversion response of various electrodes used by Cola et. Al b.Cyclicvoltammograms of various electrodes in 0.1M $K_3Fe(CN)_6/K_4Fe(CN)_6$ at 100 mV/s vs Ag/AgCl reference electrode [115]. Reproduced with permission from J. Electrochem. Soc.

Using BCNT and NCNT in an asymmetric configurations resulted in increased currents at small thermal gradients as compared to the symmetric arrangements. This current then decreased non-linearly as the temperature difference was increased. At small thermal gradients, with the BCNT at the cold side of the cell, the slower kinetics induced an accumulation of reactants at its surface. The faster kinetics at the NCNT, brought about by the increased temperature, kept the electrolyte concentration in its vicinity low. Both factors allow the redox reactions to occur rapidly until a threshold thermal gradient is reached. At this point, the ion concentration in the vicinity of the NCNT is sufficiently large to slow the kinetics and reduce the currents generated. The threshold temperature gradient is attained at lower temperatures when the NCNT is kept at the cold side of the cell because the slower kinetics at this temperature promotes accumulation of K^+ ions on the NCNT surface and leads to the "blocking" effect discussed previously.

3.2.3. Composites

The recent discovery of graphene through micromechanical exfoliation has sparked a flurry of research into its possible applications [135]. Graphene consists of a single layer of carbon atoms bonded in a hexagonal lattice. Like CNTs, its remarkable properties (charge carrier mobility of 200000 cm^2/V-s and specific surface area of 2630 m^2/g) make it an ideal candidate for electrochemical applications [136, 137]. In order to scale up graphene production, graphite is normally exfoliated in the liquid state through surfactant/solvent stablilization [138] or chemical conversion resulting in a graphene like structure known as reduced graphene oxide (RGO), shown in Figure 17 [139]. Being of the same composition as CNTs, investigation of the possibility of synthesizing composites of these two carbon materials and exploring their performance as electrode materials has been done by several research groups [83, 91, 140].

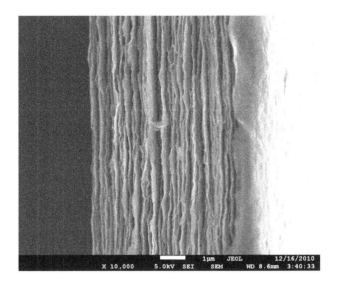

Figure 17. SEM image of the cross section of a reduced graphene oxide film

Kang et al. have shown that when composites composed of 1:1 weight RGO and P-SWNT are used in thermocells, the specific power generated (5.3 W/kg) is comparable to that of SWNTs (the experimental conditions used were discussed in section 3.2.1) [113]. It must be noted that when RGO alone is used, the specific power generated is 3.87 W/kg. However, the composite electrode produces a specific power that is only 78 % of the P-SWNT. This decrease in performance can be attributed to the large ohmic resistance observed in the RGO electrode (35.6Ω) that is 55 % higher than the P-SWNT electrode. The chemical conversion of graphite to RGO involves oxidising graphite, exfoliation and then a subsequent reduction. It is surmised that the incomplete removal of the oxygen containing functional groups is the cause of the pronounced ohmic resistance. Another reason for the poor performance of the composite electrodes is the restacking of the RGO sheets during electrode preparation, which impedes electrolyte diffusion and results in sluggish kinetics.

Optimisation of the RGO-SWNT composition for thermocell electrodes was done by Chen et. al [141]. The amount of RGO added ranged from 1 to 20 % by weight. A U-cell was used with 0.75 cm^2 electrodes separated by 10 cm, a thermal gradient of 60 °C with T_{cold}= 20 °C and 0.4 M ferri/ferrocyanide electrolyte. The optimised composite 99 % SWNT-1 % RGO generated a specific current density of 26.78 W/kg. By using large amounts of SWNTs the RGO sheets were prevented from restacking, which resulted in the appropriate nanoporosity that promoted redox mediator diffusion. The sheet like structure of the RGO provided increased pathways for electrons in the composite thus contributing to its enhanced performance. The interaction between SWNTs and RGO is clearly seen in Figure 18.

Figure 18. SEM images of reduced graphene oxide-SWNT composites

3.3. Developments in processing and fabrication of CNTs for better cell design

3.3.1. Current processing techniques for CNTs relevant to thermocells

Solvent/surfactant exfoliation

One of the major obstacles to research on CNT characterisation and application is their spontaneous aggregation brought about by attractive van der Waals interactions in both aqueous and organic solutions. The resulting aggregates or "bundles" can reach lengths of several microns and diameters of tens of nanometers. Debundling of these aggregates is essential as they have mediocre properties as compared to individual tubes. Reproducibility of results also becomes an issue when CNTs are not dispersed adequately.

Liquid phase separation is one of the simplest methods wherein stable CNT dispersions are attainable. Stable well-exfoliated CNT dispersions is achieved by appropriate selection of the solvent as forced dispersion via ultrasonication will result in agglomeration of the CNTs in a very short span of time. Selection of solvents can be based on the enthalpy of mixing per solvent volume ($\Delta \bar{H}_{mix}$) [138]:

$$\Delta \bar{H}_{mix} = \frac{2}{R_{bun}}(\delta_{NT} - \delta_{sol})^2 \varnothing \qquad (7)$$

Where R_{bun} is the radius of the dispersed nanotube bundles, δ_{NT} and δ_{sol} are square roots of the nanotube and solvent surface energies and $\varnothing$ is the nanotube volume fraction. The solubility theory states that a negative free energy of mixing ($\Delta \bar{G}_{mix}$) is indicative of a stable dispersion.

$$\Delta \bar{G}_{mix} = \Delta \bar{H}_{mix} - T \Delta \bar{S}_{mix} \qquad (8)$$

The entropy of mixing per unit volume ($\Delta \bar{S}_{mix}$) of nanotubes is generally small owing to their size and rigidity [142]. In order to realize a minimisation of $\Delta \bar{G}_{mix}$ then solvents which result in small values of ΔH_{mix} are necessary. Based on Equation 7, the most effective solvents at dispersing CNTs would be those that have a surface energy close to the nanotube surface energy (~70 mJ/m²); i.e. solvents with surface tension around 40 mJ/m² [143].

Another method to attain stable dispersions of CNTs is through the use of surfactants. Its inherent advantage over solvent dispersion was the fact that it was carried out in aqueous media, lessening its hazards and environmental impact. It relies on the principle wherein colloids are stabilized by surface charges [144]; i.e. Coulomb repulsion. Adsorption of the amphiphilic surfactant molecules onto CNTs is attained through their hydrophobic tails. This introduces a removable surface charge that creates an electric double layer around the nanotube; of which the magnitude and sign is proportional to its zeta potential [145]. This double layer provides repulsive forces that counteract the attractive van der Waals forces [146]. Selection of surfactants for CNTs dispersion depends on the size of their molecules. Low molecular weight surfactants will be able to pack tightly around the nanotube surface resulting in better stabilization [147].

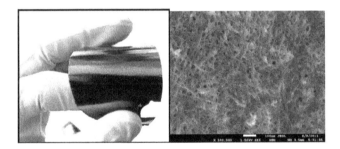

Figure 19. CNT bucky paper

Filtration of CNT dispersions results in a planar mat of randomly arranged tubes [148] that can be up to several hundred microns thick. These mats or "bucky papers" (Figure 19) have been instrumental in CNT evaluation (owing to their simple processing) as electrode materials not only for thermocells but for other electrochemical applications as well. Post-treatment of buckypapers via annealing or acid wash is essential to ensure complete removal of the solvent or surfactant used to attain the CNT dispersion [149].

Chemical vapour deposition

Chemical vapour deposition (CVD), a process that involves deposition of solids from a gas phase, has proven to be a viable method for attaining highly oriented CNTs on planar substrates (Figure 20) [150]. One of the theories behind the large degree of alignment is the reduction in free energy brought about by the van der waals interactions along the tube length inducing coordinated growth by holding the tubes together [151]. The ability to tailor the growth of CNTs in three dimensional configurations is highly advantageous in thermocell applications. This configuration promotes enhanced ion accessibility with the CNT matrix allowing larger current to be generated. The alignment of the tubes also minimises the tortuosity of the CNT electrodes which decreases the probability of forming concentration gradients within the electrode itself, leading to a decrease in the mass transfer overpotential. The wide range of substrates that can be used for CNT synthesis (metallic, carbon, etc) via CVD allows the fabrication of electrodes with a high degree of flexibility; materials that are highly desirable in thermocells [152].

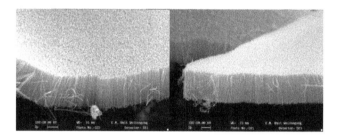

Figure 20. Aligned CNTs produced by CVD process

3.3.2. Cell design breakthroughs attained using CNTs

The development of flexible electrodes for electrochemical applications has paved the way for innovative cell designs for thermal energy conversion. CNT forests grown directly on thermocell casings, scroll electrodes, and thermocells that can be wrapped around cooling/heating pipes have been attained through flexible CNT electrodes [114].

Mark II thermocell

When inter electrode separation is decreased, larger specific power is generated as mass transfer is enhanced over shorter distances. However, this results in decreased power conversion efficiency as larger thermal energy is required to maintain a similar thermal gradient [112]. Scroll electrodes (Figure 21) can be employed to mitigate this problem. Using scrolled MWNT buck papers, each with a diameter of 0.3 cm and mass of 0.5 mg, aligned along their rolling axis inside a glass tube containing 0.4 M ferri/ferrocyanide, an electrode separation of 5 cm, T_{cold} of 5 °C and thermal gradient of 60 °C a specific power density of 1.8 W/m^2 was obtained. The power conversions efficiency of 0.24 % is an order of magnitude higher than thermocells using Pt electrodes tested under similar conditions [114]. The relative efficiency of the Mark II thermocell is 17 % higher than that obtained when using platinum electrodes, giving Φ_r of 1.4 %.

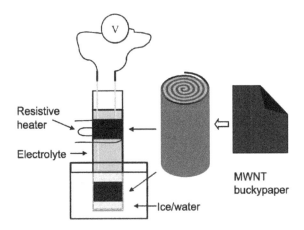

Figure 21. Mark II thermocell[114] Reprinted with permission from American Chemical Society

Coin cell

Thin coin type thermocells which could be powered by extremely low thermal gradients were developed using MWNTs and 0.4M ferri/ferrocyanide as the electrolyte (Figure 22). Coin cells fabricated using MWNT bucky paper electrodes and exposed to a thermal gradient of 45°C generated a specific power of 0.389 W/m^2 (equivalent to a normalised power density $P_{max}/\Delta T^2$ of 1.92 x 10^{-4} W/m^{2}K). Coin cells with electrodes made of MWNT forests around 100 μm tall, grown directly on the internal stainless steel surface of the packaging substrate using a trilayer catalyst (30 nm Ti, 10 nm Al, 2 nm Fe) plasma enhanced CVD method was also developed. The specific power generated at a thermal gradient of 60°C was 0.980 W/m^2,giving $P_{max}/\Delta T^2$ =2.72 x 10^{-4} W/m^{2}K. The larger normalised power density of the coin cell with MWNT forest electrodes is due to its nanotube alignment, which promotes electrolyte diffusion, and the lower thermal (0.01 cm^{2}K/W) and electrical resistance at the MWNT forest/substrate junction [153]. The thermal resistance for bucky papers is around 0.05 cm^2 K/W which leads to larger loss of thermal energy at the electrode/substrate junction and 30% less power conversion efficiency [154].

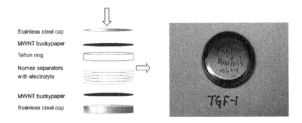

Figure 22. Coin cell for thermal energy conversion [114] Reprinted with permission from American Chemical Society

Flexible thermocell

One of the main applications of thermocells is to harvest thermal energy from automobile exhaust pipes and cooling or heating lines in industrial facilities. Flexible thermocells can be wrapped around these pipes and convert them to sources of electrical power. A flexible thermocells consisting of two MWNTbucky paper electrodes kept apart by 2 layers of NomexHT 4848 impregnated with 0.4M ferri/ferrocyanide and wrapped in a stainless steel sheetis shown in Figure 23. The cell was wrapped around a cooling pipe and a thermal gradient of 15°C was applied using a resistive heater. A specific power 0.39 W/m² was generated proving that flexible thermocells are now a possibility [114].

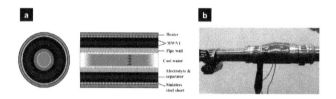

Figure 23. a) schematic b) photo of a flexible thermocell that can be wrapped around cooling/heating pipes [114] Reprinted with permission from American Chemical Society

3.4. Conclusion

Research on CNTs as electrode materials for thermogalvanic cells is still in its early stages. However, these initial results indicate that these nanocarbons are capable of generating significant amounts of power; much larger than when conventional electrodes are used. Without excluding the energy input from mechanical stirring, thermocells with platinum electrodes are able to attain a power conversion efficiency relative to a Carnot engine of 1.2% [100]. This value was surpassed with the use of CNT electrodes, reaching $\Phi_r = 1.4\%$, in a thermocell that did not utilise any mechanical stirring and relied only on convection and diffusion to cause mass transfer of reaction products. The most important breakthrough is in the area of cell design as the robustness of CNT electrodes allows them to be conformed into a variety of shapes in order to mitigate heat flow from the hot to cold side of the thermocell. Flexible thermocells are also possible; devices that can harvest heat from heating or cooling pipes. Optimisation of the porosity of these CNT electrodes is essential in order to minimise the tortuosity and to reduce the mass transport overpotential in these systems. Doping CNT electrodes can alter their electroactive surface area by up to 4-fold; this feature can be exploited by selecting the right electrolyte. The use of CNT-RGO composites has demonstrated the synergistic effect of these two materials, augmenting the power conversion efficieny of thermocells. Further developments in the field of CNT synthesis and processing will decrease the cost of these materials such that commercialisation of thermogalvanic cells may one day be realised.

4. Perspective and future developments

With energy consumption as a whole on the increase, coupled with the rapid economic development of countries such as Brazil, China, India, and Russia there will be a concerted effort to improve how energy is utilised. This expansion in industrialisation has already and will continue to lead to a further increase in the price of oil. Coupled to the rise in fossil fuel costs are drivers of an ageing energy infrastructure system and demand for a low-carbon emission economy through the use of renewable energy [155]. To help accommodate all these factors the supply and demand challenge may be addressed by tapping into otherwise wasted energy. Low grade heat, if effectively harvested can prove to be a viable source of power. Thermal converters have the potential to increase the efficiency of current energy conversion systems. Energy storage also plays a key role in providing a solution to the energy problem. Energy must be efficiently stored, when it is in excess, and released at a time of high demand. This is extremely important for renewables that are not load-following [156].

With these energy challenges and ongoing research and development, including those that have been conducted over the last decade, the awareness of the benefits of electrochemical capacitors is increasing. As the research and development into energy storage and conversion has increased, the applications of electrochemical capacitors has increased with the technology becoming more diverse meaning that systems can better be tailored /targeted for specific applications ranging from higher energy density to high power densities where fast charge / discharge efficiencies are needed [156]. The most commonly used material for supercapacitors has been activated carbon with new nanostructured materials such as carbon nanotubes and its derivatives coming to the forefront of the current fundamental research. It can be seen that the way forward in terms of trying to improve energy density and power density is in the use of CNT with composite materials such as other carbons, and conducting polymers or metal oxides in order to take advantage of the pseudocapacitative effects that these materials provide.

The research on thermogalvanic systems in the past has been generally limited to platinum electrodes [100]. This has enhanced the understanding of these electrochemical systems but has not advanced the research in terms of commercialisation due to its cost. The use of carbon nanomaterials has improved the performance of these devices immensely because of their fast transfer kinetics and large electroactive surface area and is also economically viable. A record threefold increase in power conversion efficiency (as compared to conventional systems wherein platinum is used) has been realised with the use of MWNT electrodes [114]. Flexible electrodes are now possible due to CNTs. These can be used as scroll electrodes or for thermocells that can be wrapped around pipes will make this system more versatile in terms of its possible applications. Further increase in thermocell performance may be realised with the use of CNTs-graphene composite materials.

Future development will most likely see supercapacitors and thermocells become a central part of hybrid energy storage and power delivery systems for large scale and domestic demand strategies. The integration of these two systems into one device will allow the converted waste heat to be stored then released when deemed necessary. These future advancements will not only enable better automotive and portable electronics, but they will revolutionise the fields of medicine, defence and consumer goods thus providing a step change in energy storage technology [5].

Author details

Dennis Antiohos*, Mark Romano, Jun Chen* and Joselito M. Razal

*Address all correspondence to: junc@uow.edu.au

Intelligent Polymer Research Institute, Australian Institute of Innovative Materials, Innovation Campus, University of Wollongong, Australia

References

[1] Turner, J. A. (1999). A Realizable Renewable Energy Future. *Science,* 285(5428), 687-689.

[2] Shukla, A. K. S. S., & Vijayamohanan, K. (2000). Electrochemical supercapacitors: Energy storage beyond batteries. *Current Science,* 79.

[3] Arico, A. S., et al. (2005). Nanostructured materials for advanced energy conversion and storage devices. *Nat Mater,* 4(5), 366-377.

[4] Hall, P. J., et al. (2010). Energy storage in electrochemical capacitors: designing functional materials to improve performance. *Energy & Environmental Science.*

[5] Hall, P. J., & Bain, E. J. (2008). Energy-storage technologies and electricity generation. *Energy Policy,* 36(12), 4352-4355.

[6] Murakami, T., et al. (2003). Thermoelectric Power of M-H Systems in Molten Salts and Application to M-H Thermogalvanic Cell. *Journal of The Electrochemical Society,* 150(7), A928-A932.

[7] Chung, J., et al. (2004). Toward Large-Scale Integration of Carbon Nanotubes. *Langmuir,* 20(8), 3011-3017.

[8] Frackowiak, E, et al. (2000). Supercapacitor electrodes from multiwalled carbon nanotubes. *Applied Physics Letters,* 77(15), 2421.

[9] Anantram, M.P, & Léonard, F. (2010). Physics of carbon nanotube electronic devices Rep. Prog. Phys. , 2006, 69.

[10] Pauliukaite, R., et al. Electrochemical impedance studies of chitosan-modified electrodes for application in electrochemical sensors and biosensors. *Electrochimica Acta,* 55(21), 6239.

[11] Dolatshahi-Pirouz, A., et al. (2008). Bovine serum albumin adsorption on nano-rough platinum surfaces studied by QCM-D. *Colloids and Surfaces B: Biointerfaces*, 66(1), 53.

[12] Aaron, Davies., & Yu, A. (2011). Material Advancements in Supercapacitors: From Activated Carbon to Carbon Nanotube and Graphene. *The Canadian Journal of Chemical Engineering*, 89, 1342-1357.

[13] Sherman, L. M. (2007). Carbon Nanotubes Lots of Potential--If the Price is Right. 01/05/12]; Available from:, www.ptonline.com/articles/carbon-nanotubes-lots-of-potentialif-the-price-is-right.

[14] Kierzek, K., et al. (2004). Electrochemical capacitors based on highly porous carbons prepared by KOH activation. *Electrochimica Acta*, 49(4), 515-523.

[15] Izadi-Najafabadi, A., et al. Extracting the Full Potential of Single-Walled Carbon Nanotubes as Durable Supercapacitor Electrodes Operable at 4 V with High Power and Energy Density. *Advanced Materials*.

[16] Kim, B. C., et al. (2009). Capacitive properties of RuO2 and Ru-Co mixed oxide deposited on single-walled carbon nanotubes for high-performance supercapacitors. *Synthetic Metals*, 159(13), 1389-1392.

[17] Bard, A. J., & Faulkner, L. R. (2001). Electrochemical Methods: Fundamentals and Applications. John Wiley & Sons, Inc.

[18] Yoon, S., et al. (2009). Preparation of mesoporous carbon/manganese oxide materials and its application to supercapacitor electrodes. *Journal of Non-Crystalline Solids*, 355(4-5), 252-256.

[19] Snook, G. A., Kao, P., & Best, A. S. (2011). Conducting-polymer-based supercapacitor devices and electrodes. *Journal of Power Sources*, 196(1), 1-12.

[20] Qu, Q. T., Wang, B., Yang, L. C., Shi, Y., Tian, S., & Wu, Y. P. (2008). Study on electrochemical performance of activated carbon in aqueous Li2SO4, Na2SO4 and K2SO4 electrolytes. *Electrochemistry Communications*, 10(10), 1652-1655.

[21] Bose, S., et al. (2012). Carbon-based nanostructured materials and their composites as supercapacitor electrodes. *Journal of Materials Chemistry*, 22(3), 767-784.

[22] Zhou, C. (2006). Carbon Nanotube Based Electrochemical Supercapacitors, in School of Polymer, Textile and Fiber Engineering. *Georgia Institute of Technology: Atlanta, Georgia*.

[23] Inagaki, M., Konno, H., & Tanaike, O. (2010). Carbon materials for electrochemical capacitors. *Journal of Power Sources*, 195(24), 7880-7903.

[24] Subramanian, V. R., Devan, S., & White, R. E. (2004). An approximate solution for a pseudocapacitor. *Journal of Power Sources*, 135(1-2), 361-367.

[25] Delahay, P., & Holub, K. (1968). Coupling of charging and faradaic processes: Electrode admittance for reversible processes. *Journal of Electroanalytical Chemistry*, 16(2), 131-136.

[26] Liu, T. C., Pell, W. G., & Conway, B. E. (1999). Stages in the development of thick cobalt oxide films exhibiting reversible redox behavior and pseudocapacitance. *Electrochimica Acta*, 44(17), 2829-2842.

[27] Pollak, E., Salitra, G., & Aurbach, D. (2007). Can conductivity measurements serve as a tool for assessing pseudocapacitance processes occurring on carbon electrodes? *Journal of Electroanalytical Chemistry*, 602(2), 195-202.

[28] Bradley, D. (2010). Ordered energy storage: Energy. *Materials Today*, 13(1-2), 9.

[29] Presser, V., Heon, M., & Gogotsi, Y. (2011). Carbide-Derived Carbons- From Porous Networks to Nanotubes and Graphene. *Advanced Functional Materials*, 21(5), 810-833.

[30] Zhang, L. L., & Zhao, X. S. (2009). Carbon-based materials as supercapacitor electrodes. *Chemical Society Reviews*, 38(9), 2520-2531.

[31] Battery and Energy Technologies. (2012). Available from:, http://www.mpower-uk.com/performance.htm.

[32] Conway, B. E. (1999). Electrochemical Supercapacitors: Scientific Fundamentals and Technological Applications. New York: Kluwer Acedemic / Plenum Publishers.

[33] Jong, H. J., et al. (2006). Supercapacitor Performance of Hydrous Ruthenium Oxide Electrodes Prepared by Electrophoretic Deposition. *Journal of The Electrochemical Society*, 153(2), A321-A328.

[34] Lewandowski, A., et al. (2010). Performance of carbon-carbon supercapacitors based on organic, aqueous and ionic liquid electrolytes. *Journal of Power Sources*, 195(17), 5814-5819.

[35] Kötz, R, & Carlen, M. (2000). Principles and applications of electrochemical capacitors. *Electrochimica Acta*, 45(15-16), 2483-2498.

[36] Balducci, A., et al. (2004). Ionic liquids for hybrid supercapacitors. *Electrochemistry Communications*, 6(6), 566-570.

[37] Frackowiak, E., Lota, G., & Pernak, J. (2005). Room-temperature phosphonium ionic liquids for supercapacitor application. *Applied Physics Letters*, 86(16), 164104-164103.

[38] Wei, D., & Ivaska, A. (2008). Applications of ionic liquids in electrochemical sensors. *Analytica Chimica Acta*, 607(2), 126-135.

[39] Iijima, S. (1991). Helical microtubules of graphitic carbon. *Nature*, 354(6348), 56-58.

[40] Yu, B, et al. (2006). The electrolyte switchable solubility of multi-walled carbon nanotube/ionic liquid (MWCNT/IL) hybrids. *Chemical Communications* [22], 2356-2358.

[41] Liu, J., et al. (2008). Postsynthesis microwave treatment to give high-purity multiwalled carbon nanotubes. *AIChE Journal*, 54(12), 3303-3307.

[42] Che, G., et al. (1999). Metal-Nanocluster-Filled Carbon Nanotubes:‰ Catalytic Properties and Possible Applications in Electrochemical Energy Storage and Production. *Langmuir*, 15(3), 750-758.

[43] Meyyappan, M. (2005). Carbon Nanotubes: Science & Applications. Boca Raton: CRC Press.

[44] Cai, X., Cong, H., & Liu, C. (2012). Synthesis of vertically-aligned carbon nanotubes without a catalyst by hydrogen arc discharge. *Carbon*, 50(8), 2726-2730.

[45] Liu, C, et al. (1999). Semi-continuous synthesis of single-walled carbon nanotubes by a hydrogen arc discharge method. *Carbon*, 37(11), 1865-1868.

[46] Yuge, R., et al. (2012). Characterization and field emission properties of multi-walled carbon nanotubes with fine crystallinity prepared by CO2 laser ablation. *Applied Surface Science*.

[47] Zhihua, P, et al. (2008). Investigation of the microwave absorbing mechanisms of HiPco carbon nanotubes. *Physica E: Low-dimensional Systems and Nanostructures*, 40(7), 2400-2405.

[48] Nikolaev, P. (2004). Gas-phase production of single-walled carbon nanotubes from carbon monoxide: A review of the HiPco process. *Journal of Nanoscience and Nanotechnology*, 307-316.

[49] Michael, J. Bronikowski, et al. (2001). Gas-phase production of carbon single-walled nanotubes from carbon monoxide via the HiPco process: A parametric study. *J. Vac. Sci. Technol. A.*, 19, 1800.

[50] Wang, G., Zhang, L., & Zhang, J. (2012). A review of electrode materials for electrochemical supercapacitors. *Chemical Society Reviews*, 797-828.

[51] Huang, C. W., et al. (2012). Electric double layer capacitors based on a composite electrode of activated mesophase pitch and carbon nanotubes. *Journal of Materials Chemistry*, 7314-7322.

[52] Dai, L. (2006). Carbon Nanotechnology Amsterdam: Elsevier.

[53] Deng, F., & Zheng, Q. (2009). Interaction models for effective thermal and electric conductivities of carbon nanotube composites. *Acta Mechanica Solida Sinica*, 22(1), 1-17.

[54] Obreja, V. V. N. (2008). On the performance of supercapacitors with electrodes based on carbon nanotubes and carbon activated material--A review. *Physica E: Low-dimensional Systems and Nanostructures*, 40(7), 2596-2605.

[55] Kay, Hyeok An, et al. (2001). Characterization of Supercapacitors Using Singlewalled Carbon Nanotube Electrodes. *Journal of the Korean Physical Society*, 39, S511-S517.

[56] Hu, S., Rajamani, R., & Yu, X. (2012). Flexible solid-state paper based carbon nanotube supercapacitor. *Applied Physics Letters*, 100(10).

[57] School of Pharmacy, . Research Program: University of Waterloo. 20/16/12]; Available from:, http://science.uwaterloo.ca/~foldvari/images/SWNT-MWNT.jpg.

[58] Morant, C., et al. (2012). Mo-Co catalyst nanoparticles: Comparative study between TiN and Si surfaces for single-walled carbon nanotube growth. *Thin Solid Films*, 39(520), 16-5238.

[59] Candelaria, S. L., et al. (2012). Nanostructured carbon for energy storage and conversion. *Nano Energy*, 1(2), 195-220.

[60] Chen, P. J. C., Qiu, J., & Zhou, C. (2010). *Nano Research*, 3, 594-603.

[61] Chau, T. T., et al. (2009). A review of factors that affect contact angle and implications for flotation practice. *Advances in Colloid and Interface Science*, 150(2), 106-115.

[62] Kuratani, K., , T. K., & Kuriyama, N. (2009). *Journal of Power Sources*, 189, 1284-1291.

[63] Bakhmatyuk, B. P., , B. Y. V., Grygorchak, I. I., & Micov, M. M. (2008). *Journal of Power Sources*, 180, 890-895.

[64] Pan, H., Li, J., & Feng, Y. (2010). Carbon Nanotubes for Supercapacitor. *Nanoscale Research Letters*, 5(3), 654-668.

[65] Fuertes, A. B., et al. (2005). Templated mesoporous carbons for supercapacitor application. *Electrochimica Acta*, 50(14), 2799-2805.

[66] Shen, J., et al. (2011). How carboxylic groups improve the performance of single-walled carbon nanotube electrochemical capacitors? *Energy & Environmental Science*, 4(10), 4220-4229.

[67] Wu, Z. S., et al. (2011). Doped Graphene Sheets as Anode Materials with Superhigh Rate and Large Capacity for Lithium Ion Batteries. *ACS Nano*, 5463-5471.

[68] Babel, K., & Jurewicz, K. (2002). Electrical capacitance of fibrous carbon composites in supercapacitors. *Fuel Processing Technology*, 77-78, 181-189.

[69] Béguin, F, et al. (2005). A Self-Supporting Electrode for Supercapacitors Prepared by One-Step Pyrolysis of Carbon Nanotube/Polyacrylonitrile Blends. *Advanced Materials*, 17(19), 2380-2384.

[70] Drage, T. C., et al. (2007). Preparation of carbon dioxide adsorbents from the chemical activation of urea-formaldehyde and melamine-formaldehyde resins. *Fuel*, 86(1-2), 22-31.

[71] Inagaki, N., et al. (2007). Implantation of amino functionality into amorphous carbon sheet surfaces by NH3 plasma. *Carbon*, 45(4), 797-804.

[72] Li, W., et al. (2007). Nitrogen enriched mesoporous carbon spheres obtained by a facile method and its application for electrochemical capacitor. *Electrochemistry Communications*, 9(4), 569-573.

[73] Stein, A., Wang, Z., & Fierke, M. A. (2009). Functionalization of Porous Carbon Materials with Designed Pore Architecture. *Advanced Materials*, 21(3), 265-293.

[74] Peigney, A., et al. (2001). Specific surface area of carbon nanotubes and bundles of carbon nanotubes. *Carbon*, 39(4), 507-514.

[75] Niu, J. J., et al. (2007). An approach to carbon nanotubes with high surface area and large pore volume. *Microporous and Mesoporous Materials*, 100(1-3), 1-5.

[76] Zhang, Y., et al. (2010). Preparation and electrochemical properties of nitrogen-doped multi-walled carbon nanotubes. *Materials Letters*, 65(1), 49-52.

[77] Lee, K. Y., et al. (2010). Influence of the nitrogen content on the electrochemical capacitor characteristics of vertically aligned carbon nanotubes. *Physica E: Low-dimensional Systems and Nanostructures*, 42(10), 2799-2803.

[78] Antal, A. Koósa, et al. (2010). Comparison of structural changes in nitrogen and boron-doped multi-walled carbon nanotubes. *Carbon*, 48(11), 3033-3041.

[79] Shiraishi, S., et al. (2006). Electric double layer capacitance of multi-walled carbon nanotubes and B-doping effect. *Applied Physics A: Materials Science and Processing*, 82(4), 585-591.

[80] Wang, D., , W., et al. (2008). Synthesis and Electrochemical Property of Boron-Doped Mesoporous Carbon in Supercapacitor. *Chemistry of Materials*, 20(22), 7195-7200.

[81] Gao, Y., et al. (2012). High power supercapacitor electrodes based on flexible TiC-CDC nano-felts. *Journal of Power Sources*, 201(0), 368-375.

[82] Zhou, J., et al. (2009). Mesoporous carbon spheres with uniformly penetrating channels and their use as a supercapacitor electrode material. *Materials Characterization*, 61(1), 31-38.

[83] Korenblit, Y., et al. (2010). High-Rate Electrochemical Capacitors Based on Ordered Mesoporous Silicon Carbide-Derived Carbon. *ACS Nano*, 4(3), 1337-1344.

[84] Fu, C., et al. (2011). Supercapacitor based on electropolymerized polythiophene and multi-walled carbon nanotubes composites. *Materials Chemistry and Physics*, 132(2), 596-600.

[85] Wu, Z. S., et al. (2012). Graphene/metal oxide composite electrode materials for energy storage. *Nano Energy*, 1(1), 107-131.

[86] Bhandari, S., et al. (2009). PEDOT-MWNTs commposite films. *J. Phys. Chem B*, 113, 9416-9428.

[87] Zhang, X, et al. (2011). Ultralight conducting polymer/carbon nanotube composite aerogels. *Carbon*, 49(6), 1884-1893.

[88] Crispin, X., et al. (2006). The Origin of the High Conductivity of Poly(3,4-ethylene-dioxythiophene) Poly(styrenesulfonate) (PEDOT/PSS) Plastic Electrodes. *Chemistry of Materials*, 18(18), 4354-4360.

[89] Antiohos, D., et al. (2011). Compositional effects of PEDOT-PSS/single walled carbon nanotube films on supercapacitor device performance. *Journal of Materials Chemistry*, 21(40), 15987-15994.

[90] Kim, K. S., & Park, J. S. (2011). Influence of multi-walled carbon nanotubes on the electrochemical performance of graphene nanocomposites for supercapacitor electrodes. *Electrochimica Acta*, 56(3), 1629-1635.

[91] Hu, Y., et al. (2012). Defective super-long carbon nanotubes and polypyrrole composite for high-performance supercapacitor electrodes. *Electrochimica Acta*, 60(0), 279-286.

[92] Myoungki, M., et al. (2006). Hydrous RuO_2/Carbon Black Nanocomposites with 3D Porous Structure by Novel Incipient Wetness Method for Supercapacitors. *Journal of The Electrochemical Society*, 153(2), A334-A338.

[93] Zhai, Y., et al. (2011). Carbon materials for chemical capacitive energy storage. *Advanced Materials*, 23(42), 4828-4850.

[94] Shan, Y., & Gao, L. (2007). Formation and characterization of multi-walled carbon nanotubes/Co3O4 nanocomposites for supercapacitors. Materials Chemistry and Physics (2–3): , 103, 206-210.

[95] Jayalakshmi, M., et al. (2007). Hydrothermal synthesis of SnO2-V2O5mixed oxide and electrochemical screening of carbon nano-tubes (CNT), V2O5, V2O5-CNT, and SnO2-V2O5-CNT electrodes for supercapacitor applications. *Journal of Power Sources*, 166(2), 578-583.

[96] Li, Q., et al. (2011). Structural evolution of multi-walled carbon nanotube/MnO2 composites as supercapacitor electrodes. *Electrochimica Acta*, 59(0), 548-557.

[97] Li, J. J., et al. (2012). Graphene/carbon nanotube films prepared by solution casting for electrochemical energy storage. *IEEE Transactions on Nanotechnology*, 11(1), 3-7.

[98] Dong, X., et al. (2011). The formation of a carbon nanotube-graphene oxide core-shell structure and its possible applications. *Carbon*, 49(15), 5071-5078.

[99] Wartanowicz, T. (1964). The theoretical analysis of a molten salt thermocell as a thermoelectric generator. *Advanced Energy Conversion*, 4(3), 149-158.

[100] Quickenden, T. I., & Mua, Y. (1995). A Review of Power Generation in Aqueous Thermogalvanic Cells. *Journal of The Electrochemical Society*, 142(11), 3985-3994.

[101] Kuzminskii, Y. V., Zasukha, V. A., & Kuzminskaya, G. Y. (1994). Thermoelectric effects in electrochemical systems. Nonconventional thermogalvanic cells. *Journal of Power Sources*, 52(2), 231-242.

[102] Wartsila. (2012). The world's most powerful reciprocating engine. Available from:, http://www.wartsila.com/en/engines/low-speed-engines/RT-flex96C.

[103] Dincer, I. (2002). On thermal energy storage systems and applications in buildings. *Energy and Buildings*, 34(4), 377-388.

[104] Bell, L. E. (2008). Cooling, Heating, Generating Power, and Recovering Waste Heat with Thermoelectric Systems. *Science*, 321(5895), 1457-1461.

[105] Ujihara, M., Carman, G. P., & Lee, D. G. (2007). Thermal energy harvesting device using ferromagnetic materials. *Applied Physics Letters*, 91(9), 093508-093503.

[106] Vining, C. B. (2009). An inconvenient truth about thermoelectrics. *Nat Mater*, 8(2), 83-85.

[107] Hertz, H. G., & Ratkje, S. K. (1989). Theory of Thermocells. *Journal of The Electrochemical Society*, 136(6), 1698-1704.

[108] Romano, M., et al. (2012). Novel carbon materials for thermal energy harvesting. *Journal of Thermal Analysis and Calorimetry*, 1-7.

[109] Goncalves, R., & Ikeshoji, T. (1992). Comparative studies of a thermoelectric converter by a thermogalvanic cell with a mixture of concentrated potassium ferrocyanide and potassium ferricyanide aqueous solutions at great temperature differences. *J. Braz. Chem. Soc*, 3(3), 4.

[110] Quickenden, T. I., & Vernon, C. F. (1986). Thermogalvanic conversion of heat to electricity. *Solar Energy*, 36(1), 63-72.

[111] Artjom, V. S. (1994). Theoretical study of thermogalvanic cells in steady state. *Electrochimica Acta*, 39(4), 597-609.

[112] Mua, Y., & Quickenden, T. I. (1996). Power Conversion Efficiency, Electrode Separation, and Overpotential in the Ferricyanide/Ferrocyanide Thermogalvanic Cell. *Journal of The Electrochemical Society*, 143(8), 2558-2564.

[113] Kang, T. J., et al. (2011). Electrical Power From Nanotube and Graphene Electrochemical Thermal Energy Harvesters.

[114] Hu, R., et al. (2010). Harvesting Waste Thermal Energy Using a Carbon-Nanotube-Based Thermo-Electrochemical Cell. *Nano Letters*, 10(3), 838-846.

[115] Salazar, P. F., Kumar, S., & Cola, B. A. (2012). Nitrogen- and Boron-Doped Carbon Nanotube Electrodes in a Thermo-Electrochemical Cell. *Journal of The Electrochemical Society*, 159(5), B483-B488.

[116] Quickenden, T. I., & Mua, Y. (1995). The Power Conversion Efficiencies of a Thermogalvanic Cell Operated in Three Different Orientations. *Journal of The Electrochemical Society*, 142(11), 3652-3659.

[117]   Antiohos, D., et al. (2010). Electrochemical investigation of carbon nanotube nano-web architecture in biological media. *Electrochemistry Communications*, 12(11), 1471-1474.

[118]   Landi, B. J., et al. (2009). Carbon nanotubes for lithium ion batteries. *Energy & Environmental Science*, 2(6), 638-654.

[119]   Wang, C., et al. (2003). Proton Exchange Membrane Fuel Cells with Carbon Nanotube Based Electrodes. *Nano Letters*, 4(2), 345-348.

[120]   Baughman, R. H., Zakhidov, A. A., & de Heer, W. A. (2002). Carbon Nanotubes--the Route Toward Applications. *Science*, 297(5582), 787-792.

[121]   Nugent, J. M., et al. (2001). Fast Electron Transfer Kinetics on Multiwalled Carbon Nanotube Microbundle Electrodes. *Nano Letters*, 1(2), 87-91.

[122]   Inoue, S., et al. (1998). Capillary Condensation of N2 on Multiwall Carbon Nanotubes. *The Journal of Physical Chemistry B*, 102(24), 4689-4692.

[123]   Eswaramoorthy, M., Sen, R., & Rao, C. N. R. (1999). A study of micropores in single-walled carbon nanotubes by the adsorption of gases and vapors. *Chemical Physics Letters*, 304(3-4), 207-210.

[124]   Banks, C.E, et al. (2004). Investigation of modified basal plane pyrolytic graphite electrodes: definitive evidence for the electrocatalytic properties of the ends of carbon nanotubes. *Chemical Communications* [16], 1804-1805.

[125]   Chun, K. Y., Lee, H. S., & Lee, C. J. (2009). Nitrogen doping effects on the structure behavior and the field emission performance of double-walled carbon nanotubes. *Carbon*, 47(1), 169-177.

[126]   Charlier, J. C., et al. (2002). Enhanced Electron Field Emission in B-doped Carbon Nanotubes. *Nano Letters*, 2(11), 1191-1195.

[127]   Deng, C., et al. (2009). Electrochemical detection of l-cysteine using a boron-doped carbon nanotube-modified electrode. *Electrochimica Acta*, 54(12), 3298-3302.

[128]   Panchakarla, L. S., Govindaraj, A., & Rao, C. N. R. (2007). Nitrogen- and Boron-Doped Double-Walled Carbon Nanotubes. *ACS Nano*, 1(5), 494-500.

[129]   Gong, K., et al. (2009). Nitrogen-Doped Carbon Nanotube Arrays with High Electro-catalytic Activity for Oxygen Reduction. *Science*, 323(5915), 760-764.

[130]   Terrones, M, et al. (2002). N-doping and coalescence of carbon nanotubes: synthesis and electronic properties. Applied Physics A: Materials Science & Processing , 74(3), 355-361.

[131]   Lee, Y. T., et al. (2003). Growth of Vertically Aligned Nitrogen-Doped Carbon Nanotubes: Control of the Nitrogen Content over the Temperature Range 900–1100 °C. *The Journal of Physical Chemistry B*, 107(47), 12958-12963.

[132] Banks, C. E., & Compton, R. G. (2005). Exploring the electrocatalytic sites of carbon nanotubes for NADH detection: an edge plane pyrolytic graphite electrode study. *Analyst*, 130(9), 1232-1239.

[133] Deng, C., et al. (2008). Boron-doped carbon nanotubes modified electrode for electroanalysis of NADH. *Electrochemistry Communications*, 10(6), 907-909.

[134] Strmcnik, D., et al. (2009). The role of non-covalent interactions in electrocatalytic fuel-cell reactions on platinum. *Nat Chem*, 1(6), 466-472.

[135] Novoselov, K. S., et al. (2004). Electric Field Effect in Atomically Thin Carbon Films. *Science*, 306(5696), 666-669.

[136] Bolotin, K. I., et al. (2008). Ultrahigh electron mobility in suspended graphene. *Solid State Communications*, 146(9-10), 351-355.

[137] Stoller, M. D., et al. (2008). Graphene-Based Ultracapacitors. *Nano Letters*, 8(10), 3498-3502.

[138] Coleman, J. N. (2009). Liquid-Phase Exfoliation of Nanotubes and Graphene. *Advanced Functional Materials*, 19(23), 3680-3695.

[139] Park, S., et al. (2009). Colloidal Suspensions of Highly Reduced Graphene Oxide in a Wide Variety of Organic Solvents. *Nano Letters*, 9(4), 1593-1597.

[140] Qiu, L., et al. (2010). Dispersing Carbon Nanotubes with Graphene Oxide in Water and Synergistic Effects between Graphene Derivatives. *Chemistry- A European Journal*, 16(35), 10653-10658.

[141] Romano, M. S., et al. (2012). Novel Carbon Nanomaterials for Thermal Energy Converters. in International Society of Electrochemistry 10th Spring Meeting. Perth, Australia.

[142] Bergin, S.D, et al. (2009). Multicomponent Solubility Parameters for Single-Walled Carbon Nanotube–Solvent Mixtures. *ACS Nano*, 3(8), 2340-2350.

[143] Bergin, S.D, et al. (2008). Towards Solutions of Single-Walled Carbon Nanotubes in Common Solvents. *Advanced Materials*, 20(10), 1876-1881.

[144] Israelachvili, J (1991) Intermolecular and Surface Forces. *Academic Press, London*.

[145] White, B., et al. (2007). Zeta-Potential Measurements of Surfactant-Wrapped Individual Single-Walled Carbon Nanotubes. *The Journal of Physical Chemistry C*, 111(37), 13684-13690.

[146] Hunter, R. (1994). Introduction to Modern Colloid Science. *Oxford: Oxford Science Publications.*

[147] Sun, Z., et al. (2008). Quantitative Evaluation of Surfactant-stabilized Single-walled Carbon Nanotubes: Dispersion Quality and Its Correlation with Zeta Potential. *The Journal of Physical Chemistry C*, 112(29), 10692-10699.

[148] Liu, J., et al. (1998). Fullerene Pipes. *Science*, 280(5367), 1253-1256.

[149] Geng, H. Z., et al. (2008). Absorption spectroscopy of surfactant-dispersed carbon nanotube film: Modulation of electronic structures. *Chemical Physics Letters*, 455(4-6), 275-278.

[150] Chen, J., et al. (2007). Flexible, Aligned Carbon Nanotube/Conducting Polymer Electrodes for a Lithium-Ion Battery. *Chemistry of Materials*, 19(15), 3595-3597.

[151] Wei, B. Q., et al. (2003). Assembly of Highly Organized Carbon Nanotube Architectures by Chemical Vapor Deposition. *Chemistry of Materials*, 15(8), 1598-1606.

[152] Chen, J., et al. (2008). Direct Growth of Flexible Carbon Nanotube Electrodes. *Advanced Materials*, 20(3), 566-570.

[153] Myounggu, P., et al. (2006). Effects of a carbon nanotube layer on electrical contact resistance between copper substrates. *Nanotechnology*, 17(9), 2294.

[154] Cola, B. A., et al. (2007). Photoacoustic characterization of carbon nanotube array thermal interfaces. *Journal of Applied Physics*, 101(5), 054313-054319.

[155] Wilson, I. A. G., Mc Gregor, P. G., & Hall, P. J. (2010). Energy storage in the UK electrical network: Estimation of the scale and review of technology options. *Energy Policy*, 38(8), 4099-4106.

[156] Hall, P. J. (2008). Energy storage: The route to liberation from the fossil fuel economy? *Energy Policy*, 36(12), 4363-4367.

Permissions

The contributors of this book come from diverse backgrounds, making this book a truly international effort. This book will bring forth new frontiers with its revolutionizing research information and detailed analysis of the nascent developments around the world.

We would like to thank Satoru Suzuki, for lending his expertise to make the book truly unique. He has played a crucial role in the development of this book. Without his invaluable contribution this book wouldn't have been possible. He has made vital efforts to compile up to date information on the varied aspects of this subject to make this book a valuable addition to the collection of many professionals and students.

This book was conceptualized with the vision of imparting up-to-date information and advanced data in this field. To ensure the same, a matchless editorial board was set up. Every individual on the board went through rigorous rounds of assessment to prove their worth. After which they invested a large part of their time researching and compiling the most relevant data for our readers. Conferences and sessions were held from time to time between the editorial board and the contributing authors to present the data in the most comprehensible form. The editorial team has worked tirelessly to provide valuable and valid information to help people across the globe.

Every chapter published in this book has been scrutinized by our experts. Their significance has been extensively debated. The topics covered herein carry significant findings which will fuel the growth of the discipline. They may even be implemented as practical applications or may be referred to as a beginning point for another development. Chapters in this book were first published by InTech; hereby published with permission under the Creative Commons Attribution License or equivalent.

The editorial board has been involved in producing this book since its inception. They have spent rigorous hours researching and exploring the diverse topics which have resulted in the successful publishing of this book. They have passed on their knowledge of decades through this book. To expedite this challenging task, the publisher supported the team at every step. A small team of assistant editors was also appointed to further simplify the editing procedure and attain best results for the readers.

Our editorial team has been hand-picked from every corner of the world. Their multi-ethnicity adds dynamic inputs to the discussions which result in innovative outcomes. These outcomes are then further discussed with the researchers and contributors who give their valuable feedback and opinion regarding the same. The feedback is then collaborated with the researches and they are edited in a comprehensive manner to aid the understanding of the subject.

Apart from the editorial board, the designing team has also invested a significant amount of their time in understanding the subject and creating the most relevant covers. They scrutinized every image to scout for the most suitable representation of the subject and create an appropriate cover for the book.

The publishing team has been involved in this book since its early stages. They were actively engaged in every process, be it collecting the data, connecting with the contributors or procuring relevant information. The team has been an ardent support to the editorial, designing and production team. Their endless efforts to recruit the best for this project, has resulted in the accomplishment of this book. They are a veteran in the field of academics and their pool of knowledge is as vast as their experience in printing. Their expertise and guidance has proved useful at every step. Their uncompromising quality standards have made this book an exceptional effort. Their encouragement from time to time has been an inspiration for everyone.

The publisher and the editorial board hope that this book will prove to be a valuable piece of knowledge for researchers, students, practitioners and scholars across the globe.

List of Contributors

Jing Sun and Ranran Wang
State Key Lab of High Performance Ceramics and Superfine Microstructure, Shanghai Institute of Ceramics, Chinese Academy of Sciences, China

Wei Shao, Paul Arghya, Laetitia Rodes and Satya Prakash
Biomedical Technology and Cell Therapy Research Laboratory, Department of Biomedical Engineering and Artificial cells and Organs Research Centre, Faculty of Medicine, McGill University, Canada

Mai Yiyong
Department of Chemistry, McGill University, Canada

Enid Contés-de Jesús and Carlos R. Cabrera
Department of Chemistry and NASA-URC Center for Advanced Nanoscale Materials, University of Puerto Rico, Puerto Rico

Jing Li
NASA-Ames Research Center, Moffett Field, California, USA

Tamjid Chowdhury and James F. Rohan
Tyndall National Institute, University College Cork, Lee Maltings, Cork, Ireland

Natalia V. Kamanina
Vavilov State Optical Institute, 12, Birzhevaya Line, St. Petersburg, Russia

Anne De Poulpiquet, Alexandre Ciaccafava, Saïda Benomar, Marie-Thérèse Giudici-Orticoni and Elisabeth Lojou
Bioénergétique et Ingénierie des Protéines, CNRS - AMU - Institut de Microbiologie de la Méditerranée, France

Steve F. A. Acquah, Darryl N. Ventura, Samuel E. Rustan and Harold W. Kroto
Florida State University, United States

Tawfik A. Saleh
Chemistry Department, Center of Excellence in Nanotechnology, King Fahd University of Petroleum & Minerals, Saudi Arabia

Fang-Chang Tsai, Ning Ma, Tao Jiang, Chi Zhang, Han-Wen Xiao and Gang Chang
Ministry of Education Key Laboratory for the Green Preparation and Application of Functional Materials, Faculty of Materials Science and Engineering, Hubei University, P. R. China

Lung-Chang Tsai, Chi-Min Shu, Tai-Chin Chiang, Yung-Chuan Chu, Wei-Ting Chen
Department of Safety, Health, and Environmental Engineering, National Yunlin University of Science and Technology, Douliou, Yunlin, Taiwan ROC

Hung-Chen Chang
Department of Chemical and Materials Engineering, National Chin-yi University of Technology, Taichung, Taiwan ROC

Sheng Wen
Faculty of Chemistry and Materials Science, Hubei Engineering University, P. R. China

Shih-Hsin Chen
Department of Food Science, National I-Lan University, I-Lan, Taiwan ROC

Yao-Chi Shu
Department of Cosmetic Applications & Management, Lee Ming Institute of Technology, Taipei, Taiwan ROC

Dennis Antiohos, Mark Romano, Jun Chen and Joselito M. Razal
Intelligent Polymer Research Institute, Australian Institute of Innovative Materials, Innovation Campus, University of Wollongong, Australia

Printed in the USA
CPSIA information can be obtained
at www.ICGtesting.com
JSHW011439221024
72173JS00004B/869